Lecture Notes in Computer Science 1356

Edited by G. Goos, J. Hartmanis and J. van Leeuwen

Springer
Berlin
Heidelberg
New York
Barcelona
Budapest
Hong Kong
London
Milan
Paris
Santa Clara
Singapore
Tokyo

André Danthine Christophe Diot (Eds.)

From
Multimedia Services
to Network Services

4th International COST 237 Workshop
Lisboa, Portugal, December 15-19, 1997
Proceedings

Springer

Series Editors

Gerhard Goos, Karlsruhe University, Germany

Juris Hartmanis, Cornell University, NY, USA

Jan van Leeuwen, Utrecht University, The Netherlands

Volume Editors

André Danthine
Institut Montefiore, B28
Universite De Liege, B-4000 Liege, Belgium
E-mail: Andre.Danthine@ulg.ac.be

Christophe Diot
INRIA, 2004, route des Luciolles, BP 93
F-06902 Sophia Antipolis Cedex, France
E-mail: diot@sophia.inria.fr

Cataloging-in-Publication data applied for

Die Deutsche Bibliothek - CIP-Einheitsaufnahme

From multimedia services to network services : proceedings / 4th International
COST 237 Workshop, Lisboa, Portugal, December 15 - 19, 1997. Andre Danthine ,
Christophe Diot (ed.). - Berlin ; Heidelberg ; New York ; Barcelona ; Budapest
; Hong Kong ; London ; Milan ; Paris ; Santa Clara ; Singapore ; Tokyo :
Springer, 1997
  (Lecture notes in computer science ; Vol. 1356)
  ISBN 3-540-63935-7

CR Subject Classification (1991): H.4.3, C.2, I.7.2, B.4.1, H.5.1, H.5.3

ISSN 0302-9743
ISBN 3-540-63935-7 Springer-Verlag Berlin Heidelberg New York

Typesetting: Camera-ready by author
SPIN 10661345   06/3142 – 5 4 3 2 1 0   Printed on acid-free paper

Fourth COST 237 Workshop
From Multimedia Services to Network Services
Lisboa, December 15-19,1997

Preface

COST 237 on "Multimedia Telecommunication Services" is one of the COST Actions belonging to the European COST Programme. The activities of COST 237 are organised by a Management Committee consisting of two representatives from each of the 15 countries participating in the Action and meeting twice a year to discuss advances in multimedia services and networking.

Beside the activities of its working groups which are open to the participation of all interested people, COST 237 organises each year a technical workshop and co-organises with the ACTS programme the ECMAST Conferences. This year workshop is the fourth. Lisbon follows Vienna, Copenhagen and Barcelona.

For this workshop, we privileged innovation, both in position papers and technical works. We received 24 submitted papers of which 12 were selected to compose the basis of the technical program.

To complete the technical programme, we scheduled 2 keynote speakers, 3 invited talks, and 3 working sessions.

- The Introduction keynote was given by Paulo Verissimo (From Research to Industry: how to find your way in the jungle of networked and distributed solutions) and the closing keynote by André Danthine (Traffic Management in ATM: where do we go ?).
- The invited talks were chosen to complement the technical presentations with hot topics related to multimedia networking on the Internet. Jim Kurose made a presentation on "Internet traffic measurement", Frank Kelly talked about "charging on the Internet" and Jon Crowcroft described the "lastest advance in telephony on the Internet".
- Three working sessions were scheduled to organise debates about contradictory issues. "Mobile agents" and "IP/ATM QoS" sessions were followed by a working session on the future of the COST 237 action in the context of the european research and politic on multimedia networking.

This program was arranged to give plenty of time for the discussion of hot topics.

The workshop was also introduced by two half day tutorials. Web Evolution was addressed first by Tom Greene (W3C), followed by the Multimedia Communication Environment in the afternoon by Carsten Griwodz and Lars Wolf (University of Darmstadt).

We hope you will enjoy our paper selection which is the core of these proceedings. Papers are grouped in 5 technical sessions that address the important issues of Research and Development in multimedia and network services.

Multimedia communication environments, middleware, multimedia applications, resource control, and group communication are the issues that are addressed in the technical programme.

We would like to thank for their respective contributions the Program Committee members, the reviewers, and all authors. Finally we finally thank the organisation committee for an outstanding workshop organisation in the delightful Lisbon city environment

October 1997   André Danthine and Christophe Diot

Program Committee

Patrick Baker	*HP Labs, UK*
Carlos Belo	*IST/IT, Portugal*
Jean-Yves Le Boudec	*EPFL, Suisse*
Torsten Braun	*IBM Heidelberg, Germany*
Bob Briscoe	*BT, UK*
Augusto Casaca	*INESC, Portugal*
Geoff Coulson	*Lancaster University, UK*
Jon Crowcroft	*UCL, UK*
André Danthine	*University of Liège, Belgium*
Michel Diaz	*LAAS/CNRS, France*
Wolfgang Effelsberg	*University of Mannheim, Germany*
Serge Fdida	*LIP6, France*
Domenico Ferrari	*Universita Cattolica at Piacenza, Italy*
Per Gunningberg	*University of Upsalla, Sweden*
David Hutchison	*Lancaster University, UK*
Jean-Pierre Hubaux	*EPFL, Switzerland*
Marjory Johnson	*RIACS, USA*
Helmut Leopold	*Alcatel, Austria*
Benoît Macq	*Université Catholique de Louvain, Belgium*
Jordi Domingo-Pascual	*Universita Politecnica de Catalunya, Spain*
Ramon Puigjaner	*Universita de les Illes Balears, Spain*
Bernard Plattner	*ETH Zurich, Switzerland*
Radu Popescu-Zeletin	*GMD-Fokus, Germany*
Aruna Seneviratne	*UTS, Australia*
Otto Spaniol	*Technical University Aachen, Germany*
Ralf Steinmetz	*Technical University of Darmstadt, Germany*
Harmen van As	*Vienna University of Technology, Austria*
Giorgio Ventre	*University of Napoli Federico II, Italy*
Martina Zitterbart	*Technical University Braunschweig, Germany*

List of Reviewers

Veronique Baudin
Ljubica Blazevic
Marc Boyer
Torsten Braun
Bob Briscoe
Carlos A. Carvalho Belo
Augusto Casaca
Christophe Chassot
Geoff Coulson
Jon Crowcroft
C. Cseh
Michel Diaz
F. Dietrich
Christophe Diot
Jordi Domingo-Pascual
Christopher Edwards
Wolfgang Effelsberg
Serge Fdida
Stephan Fischer
Xavier Garcia Adanez
Per Gunningberg
Marjory Johnson
Dogan Kesdogan
Bjorn Knutsson
Andreas Kraft
Axel Kuepper
Jean-Yves Le Boudec
Helmut Leopold
Thomas Luckenbach
Benoit Macq
Bernhard Plattner
Ramon Puigjaner
Marko Schuba
Gerd Schuermann
Aruna Seneviratne
Jerome Tassel
Dirk Thissen
Dirk Trossen
Harmen R. van As
Daniel G. Waddington
Thomas Walter
Martina Zitterbart

Steering Committee

Theodoros Bozios	*Intracom, Greece*
André Danthine	*University of Liege, Belgium*
Jordi Domingo-Pascual	*University Politecnica de Catalunya, Spain*
Wolfgang Effelsberg	*University of Mannheim, Germany*
Serge Fdida	*LIP6, France*
Domenico Ferrari	*University Cattolica at Piacenza, Italy*
Josi Guimarces	*ADETTI, Portugal*
Jean-Pierre Hubaux	*EPFL, Switzerland*
David Hutchison	*Lancaster University, UK*
Villy Iversen	*Technical University of Denmark, Denmark*
Borka Jerman-Blazic	*Institute Jozef Stefan, Slovenia*
Helmut Leopold	*Alcatel, Austria*
Radu Popescu-Zeletin	*GMD-Fokus, Germany*
Sandor Stefler	*PKI, Hungary*
Giorgio Ventre	*University of Napoli, Italy*
Branka Zovko-Cihlar	*University of Zagreb, Croacia*

Organising Committee

José Guimarães	*ADETTI/ISCTE, Portugal*
Carlos Serrão	*ADETTI, Portugal*
António Almeida	*IST/IT, Portugal*
Carlos Belo	*IST/IT, Portugal*
Ciaran O'Colmain	*Norcontel, Ireland*

The Fourth COST 237 Workshop is organized by ADETTI/ISCTE with cooperation from IT/IST.

ADETTI - Associação para o Desenvolvimento das Telecomunicações e Tecnologias de Informática
Edifício ISCTE
Av. das Forças Armadas
1600 LISBOA
PORTUGAL

The Workshop is sponsored by:

COST - European Cooperation in the field of Scientific and Technical Research
ISCTE - Instituto Superior de Ciências do Trabalho e da Empresa
JNICT - Junta Nacional de Investigação Científica e Tecnológica

Operator Based Composition of Structured Multimedia Presentations

Chérif Keramane, Andrzej Duda

LSR-IMAG,
BP 72,
38402 St. Martin Cedex, France

Abstract. We are interested in a new class of multimedia presentations
that include rich dynamic interactive scenarios. Such scenarios integrate
dynamic user control over the flow of presentation, advanced processing of media content, and diverse sources of media streams such as live
feeds or teleconferencing streams. We propose a new model for structured
temporal composition of interactive dynamic multimedia presentations.
It extends the notion of basic media segments to include executable code,
live feeds, and links. In this way, we can take into account user interactions, content-sensitivity, new interesting sources of multimedia data, and
provide support for sharing and reuse. These new media segment types
are integrated in a seamless way within our temporal composition model.
The model is based on *Interval Expressions* that involve media segments
of unknown duration. We define a set of operators that express causal
relations between intervals. Operators take time intervals as arguments
and yield another interval as a result. They can be used to form nested
interval expressions allowing specification of temporal compositions in a
well-structured way. *Interval Expressions* provide a means of encapsulation and structuring: compound encapsulated intervals can be specified
in terms of elementary media objects as building blocks. We address
the temporal consistency problem—*Interval Expressions* guarantees the
absence of temporal inconsistencies by construction.
Keywords : Temporal Representation of Multimedia, Media Synchronization

1 Introduction

Multimedia representation and computing have made significant progress during
last several years. However, the advances have not led to satisfactory proposals
for rich dynamic adaptive story environments [3]. Existing models of multimedia
presentations allow static composition of different media having inherent temporal behavior such as video and audio. Temporal composition defines synchronization among media segments according to some static temporal scenario. Usually,
a presentation is played back linearly with little (simple Temporal Access Control functions) or no user control over the flow of presentation. In opposition
to the traditional multimedia presentations, we are interested in a new class of
multimedia presentations that include rich dynamic interactive scenarios. Such
scenarios integrate dynamic user control over the flow of presentation, advanced

processing of media content, and diverse sources of media streams such as live feeds or teleconferencing streams.

To be able to compose dynamic interactive multimedia presentations, we integrate several new features. First, we want to extend the notion of media to executable code. For example, some parts of a multimedia presentation can be composed of operations that analyze the content of media segments, process online multimedia streams or evaluate some queries on a multimedia collection. In that way, the outcome of the operations can determine the flow of a presentation. This content-sensitivity becomes an important new characteristic of media segments—a segment can be made sensitive to its content and the flow of a presentation can change according to the detection of some content features. Second, we want to take into account user interactions. The user can control the flow of presentation, choose different paths, and even jump from one presentation to another. Third, we want to include live feeds as components of multimedia presentations. With the advent of television broadcasting coming into a computer, multimedia presentations may take advantage of rich sources of streams coming from the live feeds such as a satelite dish or a teleconferencing network connection. An interesting example of a live media stream is the Wearable Wireless Webcam experiment[1]. Their integration requires new functions such as content analysis and feature detection.

There is a one important consequence of this extended, general view of media segments: the duration of media segments is *a priori* unknown, for example, a segment can represent a live feed or a user interaction can stop playback for some period of time. Many existing models for temporal composition of multimedia data such as the timeline or the Allen relations [1] cannot deal with such media segments because they require the knowledge of interval durations.

Moreover, existing models do not sufficiently take into account user interactions. For example, Hirzalla *et al.* propose an extended timeline model that switches to sub-timelines according to user choices [12]. MHEG standard provides mechanisms for specifying user input events and associating reactions with them [18]. LMDM defines a fine-grain scripting language for temporal composition that includes user interactions [21]. However, integrating user interactions with intervals of unknown duration calls for another approach.

Another problem with existing multimedia representations is related to temporal inconsistencies. A presentation is temporally inconsistent if some contradictory relations are specified for media segments. Authoring complex presentations requires computationally expensive algorithms for verifying temporal consistency. The verification is a required feature, but it would be better to use a model that guarantees the absence of inconsistencies.

The last issue concerns sharing and reusing of media content. We can think of multimedia authoring as a complex process of specifying pieces of a puzzle in which the pieces come from a large collection of media segments accessed by content. The author looks for interesting media segments or parts of existing presentations and combines them together to form a new presentation.

---

[1] follow http://media.mit.edu to Steve Mann

This authoring method requires support for structuring and hierarchical composition. It should be possible to define complex multimedia presentations in a well-structured manner by means of encapsulation and nesting of structured entities. For example, multimedia modeling based on algebraic construction of encapsulated presentations [25] was designed for supporting sharing and reusing of media content.

For all these reasons, we need a new multimedia representation that allows temporal composition of media segments of unknown duration, takes into account user interactions, addresses the consistency problem, and provides support for sharing and reuse. In this paper, we define a new model called *Interval Expressions* that addresses all these requirements. The model is based on intervals of unknown duration and on operators that express causal relations between intervals. Operators take time intervals as arguments and yield another interval as a result. They can be used to form nested interval expressions allowing specification of temporal compositions in a well-structured way. *Interval Expressions* provide a means of encapsulation and structuring: compound encapsulated intervals can be specified in terms of elementary media objects as building blocks.

We build upon our previous work on *Algebraic Video* [25] that uses a set of basic operations on video segments to create a desired video stream. The video algebra consisted of operations for temporally and spatially combining video segments, and for attaching attributes to these segments. Algebraic Video focused on content-based access to video libraries with query based discovery and algebraic combination of video segments of interest. We adopt the approach consisting of defining a set of operators to form nested expressions.

In the remainder of this paper, we present existing models (Section 2), present the proposed composition model (Section 3), give some examples (Section 4), and outline conclusions (Section 5).

2 Existing models

Existing temporal models for multimedia can be divided into two classes: *point-based* and *interval-based* [23]. In point-based models, the elementary units are points in a time space. Each event in the model has its associated time point. The time points arranged according to some relations such as *precede, simultaneous* or *after* form complex multimedia presentations. An example of the point-based approach is *timeline*, in which media objects are placed on several time axes called *tracks*, one per each media type. All events such as the beginning or the end of a segment are totally ordered on the timeline. The timeline model is applied in HyTime [13] and in several other propositions [10], [7]. The model is well suited for temporal composition of media segments of known durations, however it falls short for unknown durations.

Other authors have proposed to use relations between interval end points for temporal composition of multimedia (temporal point nets [2], MME [6]). The temporal point nets can deal with intervals of unknown duration, but their use is difficult and results in complicated, unstructured graphs. In addition to that, their use may lead to an inconsistent specification in which contradictory conditions are specified for intervals. In this case, a verification algorithm must check for temporal inconsistency [2].

Interval-based models consider elementary media entities as time intervals ordered according to some relations. Existing models are mainly based on the relations defined by Allen for expressing the knowledge about time [1][2]. Giving any two time intervals, they can be arranged according to seven relations : *before, meets, overlaps, finishes, during, starts, equals*. A relation expresses the inequalities between the end points of intervals. If the Allen relations are used for multimedia composition, then the duration of intervals may vary only within the limits defined by the inequalities of a given relation. This drawback makes the Allen relations not suitable for specifying composition of intervals with unknown duration. Consider for example an existing relation *before* between intervals a and b (see Figure 1). When we increase the duration of interval a, the relation changes from *before* to *during* passing through intermediate relations *meets, overlaps*, and *finishes*.

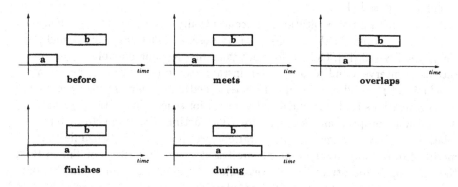

Fig. 1. Relations change when the interval duration is modified.

Another problem with the Allen relations is their descriptive character—they allow expression of an existing, *a posteriori* arrangement of intervals, but they do not express any causal or functional relation between intervals. For example, relation *meets* is ambiguous—it only states that the end of the first interval coincides with the end of the second one, but it does not say whether the first interval *starts* the second one, whether the second interval *stops* the first one or whether it is a pure coincidence (see Figure 2). So, the Allen relations can be useful for characterizing an existing, *instantiated* presentation (a presentation for which all start and termination instants of media segments are known).

The third problem with the Allen relations is related to inconsistent specifications that can be introduced to a multimedia presentation. Detecting incon-

---

[2] although the relations have already appeared in the literature much before the Allen article [11], we follow the familiar multimedia terminology that attributes the relations to Allen

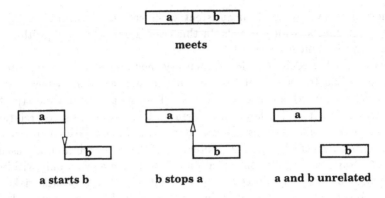

Fig. 2. *meets* represents temporal coincidence not a functional relationship.

sistent specification requires algorithms of complexity $[O(N^2)]$, where N is the number of intervals [1]. Another algorithm for detecting inconsistencies has been recently proposed [15].

Many existing representations are conceptually equivalent to the Allen relations. For example, Little and Ghafoor propose a model based on Timed Petri Nets—OCPN [16, 17]. OCPN does not take into account possible unknown durations of intervals and to prepare an instantiated presentation (a presentation in which all interval end points are determined), we must traverse the tree of interval relations to determine deadlines used for scheduling of the presentation. The authors propose an algorithm for constructing a schedule from temporal relations, however, when interval durations are modified, the schedule must be modified too. Many other proposals are based on other variants of Timed Perti Nets [9, 26]. Time Stream Perti Nets allow representing a minimum, a nominal and a maximum duration of time intervals. They provide a useful tool for modeling fine-grain temporal sychronization of media, for example lip-sync in videoconfcrencing streams [4, 22, 5, 19].

Schloss and Wynblatt define the *Layered Multimedia Data Model* composed of several layers: Data Definition, Data Manipulation, Data Presentation, and Control Layer [21]. The model includes media objects and scripted objects. Scripted objects do not contain media data, but instead they contain a set of instructions for generating data. Data Manipulation Layer defines temporal specification that relates *events* by means of an event calculus. The calculus provides operators for sequencing and temporally overlaying the occurrences of objects in an event. The model defines a sort of a fine-grain scripting language for temporal composition that includes user interactions.

FLIPS is a model for specifying coarse synchronization for flexible presentations [20]. It extends temporal point nets by defining two conditions between events (interval end points): a *barrier* (event A no earlier than event B) and an *enabler* (event A no later than event B). The model deals with interval of unknown duration, but it has the same drawback as temporal nets do: a presentation is a complex, unstructured graph and temporal consistency must be verified. An interesting feature of the model are user interactions allowing *forward* and *backward* skips in a complex presentation.

TIEMPO [24] considers integration of user interactions in a previously developed temporal composition model [23]. A fixed set of user interactions is defined and it includes either selections or Temporal Access Control functions such as for example *pause*, *faster* or *backward*. The model also proposes hierarchical grouping by means of macros.

King proposes a different formalism based on an interval temporal logic [14]. He shows how the Allen relations can be expressed using temporal logic formula. Although his formalism has solid mathematical bases, composition of multimedia presentations using declarative formula is awkward—logic formula do not correspond to the mental image that an author uses during conception. Moreover, to be useful, the formalism must be supported by a consistency checker and an interpreter to execute a given temporal specification.

Unlike previous models, *Interval Expressions* offer a different paradigm that allow specification in a structured manner of functional relations between intervals of unknown duration. We extend the notion of basic media segments to include executable code, live feeds, and links. For temporal composition, we explore the same approach as *Algebraic Video* [25] that used a set of basic operations on video segments. The video algebra defined three temporal operators for combining video segments: *concatenation*, *parallel* and *parallel-end*. Other operators such as *union*, were related to attributes and content-based access. Temporal composition based on algebraic construction of nested, encapsulated expressions supports sharing and reusing of media content. Our previous work dealt with temporal operators for intervals in the context of multimedia databases [8].

3 Temporal composition model based on *Interval Expressions*

This section presents our new temporal composition model based on *Interval Expressions*. First, we define basic media segments that can be combined together to form complex presentations. Then, we analyze possible relationships between any two time intervals and finally, we define formally operators representing the relationships.

3.1 Basic media segments

The elementary entities of our temporal model are *media segments* that have inherent temporal behavior. They may be traditional multimedia objects containing video, audio, animations, images or text as well as new special objects such as live media streams, executable code or links to other presentations. For temporal composition, media segments are viewed as *time intervals* independently of their content or other characteristics. Time interval a is defined by its end points ($\underline{a} \leq \bar{a}$) as $a = \{t \mid \underline{a} \leq t \leq \bar{a}\}$. The duration of interval a is $\bar{a} - \underline{a}$

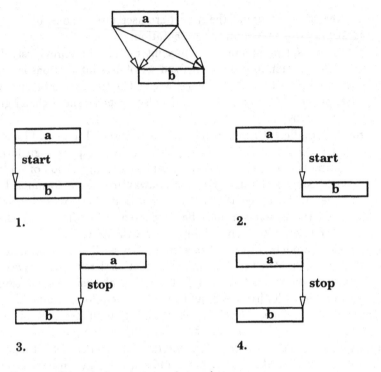

Fig. 3. Four possible relations between intervals.

and it depends on the type of media segments. We define the following types of media segments:

- Set M of traditional multimedia objects such as video, audio, animations, images, text, or pure delay. Segments of type M are time intervals with duration that is either *implied*, i.e. determined by the rate of their capture (for example, a video clip composed of 300 frames captured at 30 frames per second will have duration 10s), or *synthetic*, i.e. a delay is associated with a given segment (for example, display an image for 10s). The duration can be modified by the user during playback. The instant of activation of a media segment is determined by the presentation that uses it. The instant of the end is determined by the duration of the segment.

- Set S of live media streams such as live feeds, teleconferencing streams, or even complex presentations that arrives on a network connection. The duration of segments, the instants of their activation, and the instants of their end are *a priori* unknown and they are determined by the source of data asynchronously.

- Set P of programs. When activated, a program begins its execution. A program can take into account user interactions. For example, a program waits for a user action such as a click and terminates when the action is done. The instant of activation of a media segment is determined by the presentation that uses it. The instant of the end corresponds to the end of the execution, so that the duration of a media segment depends on the program itself.

- Set L of links. A link refers to another media segment that will be activated when a link is activated. The duration of a link is zero. The instant of its activation is determined by the presentation that uses it.
- $\emptyset$ as an empty segments.

Given these different types, we define the basic set I_{base} of media segments as follows:

$$I_{base} = M \cup S \cup P \cup L \cup \emptyset \tag{1}$$

It represents all media segments that can be used as arguments in interval operators described below.

3.2 Analysis of interval relations

Having defined basic media segments, we want to define a temporal composition model that uses them. Let us consider relationships between any two media segments considered as time intervals (see Figure 3). Each end point of an interval may perform an action on any of the end points of another interval: the action aimed at the beginning of an interval corresponds to *start* or activation and the action aimed at the end of an interval corresponds to *stop* or termination. The granularity of media segments depends on the way they are used in presentations. If it is necessary to perform an action in the middle of an interval, we can divide it into two intervals and attach the action to some of the end points. The arrows in the figure represent such actions (note that the actions represent *causal* relations that have real system interpretation which is not the case for the Allen relations).

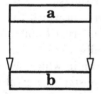 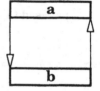

Fig. 4. Two relations for equality.

There are four possible relations with a single action involving a and b (shown in Figure 3) and four inverse relations involving b and a. In addition, we can consider relations with more than one action. Many of such relations are trivial, for example relations $\underline{a} \longrightarrow \underline{b}$ and $\underline{a} \longrightarrow \bar{b}$ make the duration of b null. There are two non trivial relations concerned with equality of interval durations (see Figure 4). In the first relation, a *starts* b and a *stops* b. In the second one, a *starts* b and b *stops* a. The distinction between the two relations is important, because

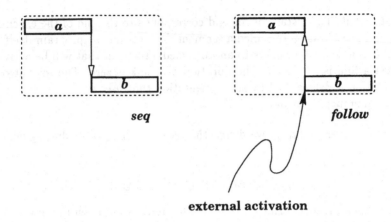

Fig. 5. Sequential temporal composition operators

any of the intervals may be of unknown duration, for example, the duration can be controlled by the user or can be determined by a live media stream externally.

Given these possible relations between intervals, we define a high-level model that provides a means for encapsulation and structuring. It is based on the following functional operators defined below (for each operator, we give its semantics and the definition of the result interval):

$$\texttt{seq}\,(a,\,b) \mapsto (\underline{a}, \bar{b})$$ defines composition in which the end of interval a *starts* interval b.

$$\texttt{follow}\,(a,\,b) \mapsto (\underline{a}, \bar{b})$$ defines composition in which the beginning of interval b *stops* interval a; interval b is activated externally.

$$\texttt{par-begin}\,(a,\,b) \mapsto (\underline{a}, \bar{a})$$ defines composition in which the beginning of interval a *starts* interval b.

$$\texttt{par-end}\,(a,\,b) \mapsto (\underline{a}, \bar{a})$$ defines composition in which the end of interval a *stops* interval b; interval b is activated externally.

$$\texttt{par-min}\,(a,\,b) \mapsto (\underline{a}, \min(\bar{a}, \bar{b}))$$ defines composition in which the beginning of interval a *starts* interval b; the result interval is stopped when the first of the two interval terminates.

$$\texttt{par-max}\,(a,\,b) \mapsto (\underline{a}, \max(\bar{a}, \bar{b}))$$ composition in which the beginning of interval a *starts* interval b; the result interval is stopped when the last of the two interval terminates.

$$\texttt{equal}\,(a,\,b) \mapsto (\underline{a}, \bar{a})$$ defines composition in which interval a *starts* and *stops* interval b.

$$\texttt{ident}\,(a,\,b) \mapsto (\underline{a}, \bar{b})$$ defines composition in which the beginning of interval a *starts* b and the end of interval b *stops* a.

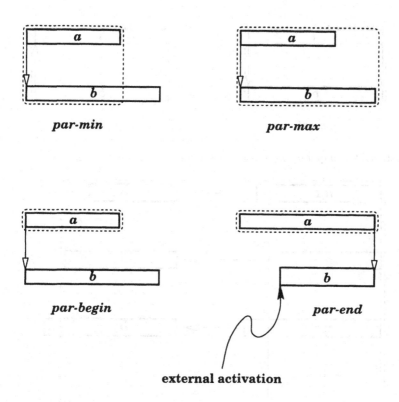

Fig. 6. Parallel temporal composition operators

`alternative` $((a_1, b_1), \ldots, (a_n, b_n)) \mapsto (\underline{a}_1, \bar{c})$ defines the composition in which intervals a_i, $i = 1, \ldots, n$ are *started* in parallel; the end of the first of them *stops* all parallel intervals and *starts* its associated interval b_i: $\bar{c} = \bar{b}_j$ where $\bar{a}_j = \min_{i=1,\ldots,n}(\bar{a}_i)$.

`loop` $a \mapsto (\underline{a}, \infty)$ defines composition in which the interval a is repeated for ever and can only be stopped by other intervals.

The operators are represented graphically in Figures 5, 6, 7, 8. Dashed boxes show the result interval for each operator (encapsulation boundary). External activations of live media streams are represented by a curved arrow.

An operator takes one, two, or more intervals as arguments and returns an interval as a result. `alternative` operator defines a correspondence between two sets of intervals: each interval of the first set has its corresponding interval in the second set. When `alternative` begins, all intervals of the first set behave as `par-min` with n arguments: they are activated in parallel and the shortest interval terminates all parallel intervals. Then, it selects its corresponding interval from the second set. The operator allows composition of conditional presenta-

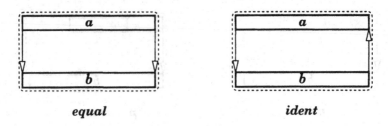

equal   *ident*

Fig. 7. Equality temporal composition operators

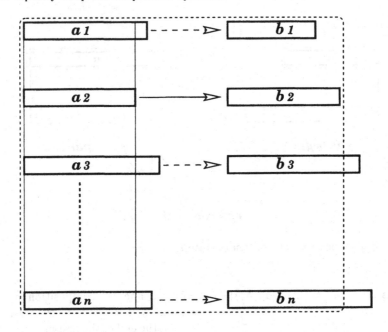

alternative

Fig. 8. Alternative temporal composition operator

tions in which the user can choose the flow of presentation.

Complex temporal multimedia compositions can be built by nesting operators. The set I of interval expressions that define a temporal multimedia composition is build from the set of media segments I_{base} by induction according to the following rule:

$$I_0 = I_{base} \tag{2}$$

$$I_{n+1} = \bigcup_{f \in \mathcal{F}} f(I_n^{ar(f)}) \cup alternative(\mathcal{P}(I_n \times I_n)) \tag{3}$$

$$I = \bigcup_{n=0}^{\infty} I_n \tag{4}$$

where $\mathcal{P}(X)$ is the set of all subsets of set X, $\mathcal{F}$ is the set of the operators except `alternative`, and $ar(f)$ the arity of operator f.

Any media segment uses as an argument of operators can be of link type. For example, Figure 9 shows two presentations and a link segment used for skipping from one presentation to another.

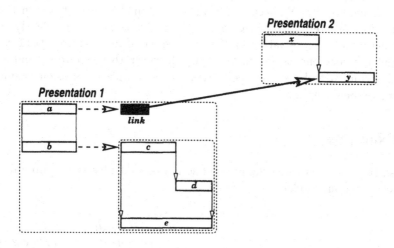

Fig. 9. Example of a link segment

The operators can be used to form nested *Interval Expressions* allowing specification of temporal compositions by means of encapsulation in a well-structured way. Temporal composition consists of considering intervals as building blocks that can be put together, encapsulated and nested. Nesting allows building more complex hierarchical well-structured presentations and encourages sharing and reuse of parts of presentations.

The structure defined by an interval expression is invariant with respect to the duration of component intervals. This means that an expression defines the temporal structure of a presentation. Its playback may result in many different temporal schedules depending on the duration of media segments. However, relations between segments are always maintained according to the temporal structure.

By construction, our model guarantees the absence of temporal inconsistencies. As each result interval is temporally consistent with the arguments of an operator, nesting preserves temporal consistency.

Forming interval expressions provides encapsulation boundaries. From the external point of view, the result interval has well defined end points so that it can be used in other expressions. From the internal point of view, all operators except `follow` and `par-end` define the internal temporal behavior of the result interval completely—the activation of the beginning end point in the result interval implies the activation of all its internal components. In this way, the result interval forms an autonomous entity separated from outside. For example, when *a starts b* (Figure 3), the result interval is defined by the end points $(\underline{a}, \bar{b})$. The activation of the result interval implies activation of *a* that in turn causes activation of *b*.

Operators `follow` and `par-end` require external activation. For example, relations 3 and 4 (Figure 3) in which *a stops b* can only be encapsulated, if we assume that interval *b* is of type S (or is a complex expression that begins with a media segment of type S), because for this type of media segments the instant of its activation is determined by the source of a media stream independently of the initial activation of interval *a*. So, the result interval for operators `follow` and `par-end` provides encapsulation boundary, however the operators require the second argument to be a live media stream or a complex expression beginning with a live media stream.

4 Examples

This section presents three examples of scenarios that illustrate temporal composition using our model.

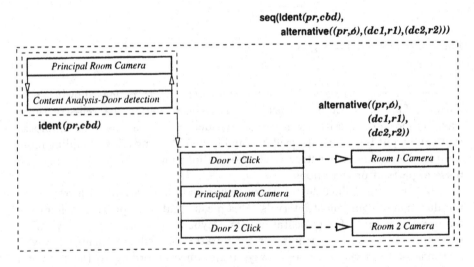

Fig. 10. Example: Guided Tour of a museum

The first example concerns a *Guided Tour* of a museum (see Figure 10). The tour begins with an on-line live video stream coming from a camera placed in a principal room. A program of content analysis of the video runs in parallel with the video stream and controls the stream by means of the `ident` operator. When two doors are detected, the program stops and activates two other programs that detect user interactions—when the user clicks on one of the images of the doors, the click terminates the program corresponding to that door. The `alternative` operator associates different media segments to each click detection program.

The media segments are live streams coming from the cameras showing rooms behind the doors. The user can see the room behind the chosen door.

Another example is the *Interactive News* (see Figure 11). This presentation allows the user to switch between two live video streams interactively. At the beginning, the video streams are active, but only the first stream is displayed. Three programs control the presentation of the streams: one that detects *Sport*, another one that detects *Weather*, and the last one that waits for a user click. Three alternative scenarios are associated with the programs. If *Sport* appears in the second stream, the presentations are merged: the second stream presented in a small window overlays the first stream. If *Weather* appears, the second flow is also overlaid on the first stream. At any time, the user can preview the other stream by clicking. The loop operator allows returning to the initial state.

The last example shows a presentation that applies *Automatic Natural Language Translation* to multimedia data (see Figure 12). A phoneme detection program, which runs in a loop, processes an on-line speech audio stream. Another program analyzes the syntax of the detected phonemes and controls the loop by means of the *ident* operator. When it detects the end of a sentence, it activates a program that translates the sentence. When the translation is finished, the program activates a synthesis audio program that generates the speech for the translated sentence in the target language. The outer loop repeats these operations continuously.

The examples show expressing power of our model and its capacity to specify complex dynamic scenarios. The scenarios are rich and interactive, because executable code and user interactions are integrated within the same temporal model.

5 Conclusions

We have presented a new model for structured temporal composition of interactive dynamic multimedia presentations. Our work differs from earlier works in several major aspects. We extend the notion of basic media segments to include executable code, live feeds, and links. In this way, we can take into account user interactions, content-sensitivity, new interesting sources of multimedia data, and provide support for sharing and reuse. These new media segment types are integrated in a seamless way within our temporal composition model. The model is based on *Interval Expressions* that involve media segments of unknown duration. We define a set of operators that express causal relations between intervals. Operators take time intervals as arguments and yield another interval as a result. They can be used to form nested interval expressions allowing specification of temporal compositions in a well-structured way. *Interval Expressions* provide a means of encapsulation and structuring: compound encapsulated intervals can be specified in terms of elementary media objects as building blocks. We address the temporal consistency problem—*Interval Expressions* guarantees the absence of temporal inconsistencies by construction.

loop(alternative(*(f1,ó)*,
                 (f2,ó),
                 (ds, equal *(f1,*ident*(f2,ds))*,
                 *(dw,*equal *(f1,*ident*(f2,dw))*,
                 (dc, equal *(f1,*ident*(f2,dc))))*

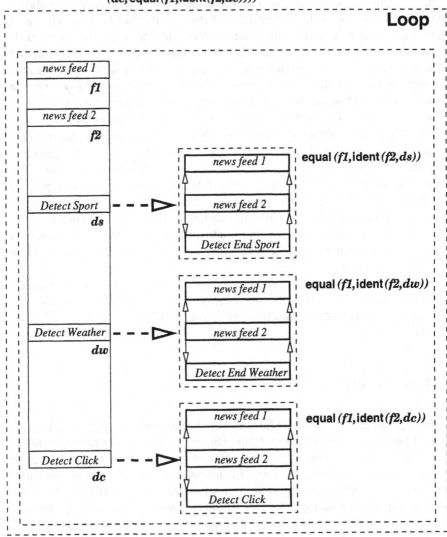

Fig. 11. Example: Interactive News

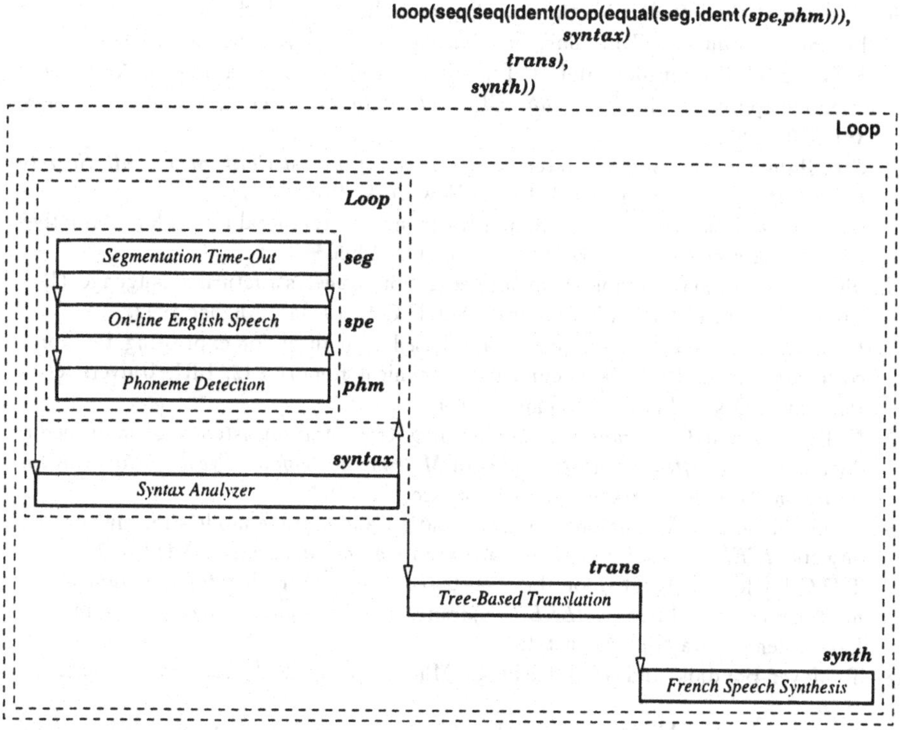

loop(seq(seq(ident(loop(equal(seg,ident(spe,phm))),
syntax)
trans),
synth))
Loop
Loop
Segmentation Time-Out
seg
On-line English Speech
spe
Phoneme Detection
phm
syntax
Syntax Analyzer
trans
Tree-Based Translation
synth
French Speech Synthesis

Fig. 12. Example: Automatic Natural Language Translation

References

1. J. F. Allen. Maintaining knowledge about temporal intervals. *Communications of the ACM*, 26(11), November 1983.
2. M. C. Buchanan and P. T. Zellweger. Automatic temporal layout mechanisms. In *Proc. First ACM International Conference on Multimedia.*, pages 341–350, Anaheim, CA, August 1993.
3. G. Davenport. Still seeking : Signposts of things to come. *IEEE Multimedia*, 2(3), 1995.
4. M. Diaz and P. Sénac. Time Stream Perti Nets, a model for multimedia stream synchronization. In *Proc. Int. Conference on Multimedia Modeling*, Singapore, 1993.
5. M. Diaz and P. Sénac. Time Stream Petri Nets, a model for timed multimedia information. In *Proc. 15th Int Conference on Application and Theory of Petri Nets*, Zaragoza, Spain, June 1994.
6. D. Dingeldein. Modeling multimedia-objects with MME. In *Proc. Eurographics Workshop on Object-Oriented Graphics*, Sintra, Portugal, May 1994.
7. G. D. Drapeau. Synchronization in the MAEstro multimedia authoring environment. In *Proc. First ACM International Conference on Multimedia.*, pages 331–339, Anaheim, CA, August 1993.
8. A. Duda and C. Keramane. Structured temporal composition of multimedia data. In *Proc. International Workshop on Multi-media Database Management Systems*, Blue Mountain Lake, NY, August 1995.

9. F. Fabre et al. A toolkit for the modeling of multimedia synchronization scenarios. In *Proc. Second Int. Conference on Multimedia Modeling*, Singapore, 1995.

10. S. Gibbs, C. Breiteneder, and D. Tsichritzis. Audio/Video databases: An object-oriented approach. In *Proc. 9th IEEE Int. Data Engineering Conference*, pages 381–390, 1993.

11. C.L. Hamblin. Instants and intervals. In *Proc. of the 1st Conf. of the Intl. Society for the Study of Time*, pages 324–331, New York, 1972.

12. N. Hirzalla, Ben Falchuk, and Ahmad Karmouch. A temporal model for interactive multimedia scenarios. *IEEE Multimedia*, 2(3), 1995.

13. ISO. Information technology hypermedia/time-based structuring language (Hy-Time). *ISO International Standard*, (ISO/IEC IS 10744), August 1992.

14. P. R. King. Towards a temporal logic based formalism for expressing temporal constraints in multimedia documents. Technical Report 942, LRI, Université de Paris-Sud, Orsay, France, December 1994.

15. N. Layaida and C. Keramane. Maintaining temporal consistency of multimedia documents. In *Effective Abstractions in Multimedia Layout, Presentation and Interaction ACM 95 Workshop*, San Francisco, CA, 1995.

16. T.D.C Little and A. Ghafoor. Synchronization and storage models for multimedia objects. *IEEE Journal on Selected Areas in Communication*, 8(3):413–427, 1990.

17. T.D.C Little and A. Ghafoor. Interval-based temporal models for time-dependent multimedia data. In *Proc. IEEE Conference on Data and Knowledge Engineering*, Lake Buena Vista, FL, August 1993.

18. T. Meyer-Boudnik and W. Effelsberg. Mheg explained. *IEEE Multimedia*, 2(1), 1995.

19. P. Owezarski and M. Diaz. Hierarchy of Time Stream Petri Nets models in generic videoconferences. In *Proc. Mutimedia and Concurrency Workshop*, Toulouse, Framce, June 1997.

20. J. Schepf, J. D. Konstan, and D. Du. Doing flips : Flexible interactive presentation synchronization. In *Proc. 1995 International Conference on Multimedia Computing and System*, Washington, DC, May 1995.

21. G. A. Schloss and M. J. Wynblatt. Building temporal structures in a layered multimedia data model. In *Proc. Second ACM International Conference on Multimedia.*, pages 271–278, San Francisco, CA, 1994.

22. P. Sénac et al. Toward a formal specification of multimedia synchronization. *Annals of telecommunications*, May/June 1994.

23. T. Wahl and K. Rothermel. Representing time in multimedia systems. In *Proc. IEEE International Conference on Multimedia Computing and Systems.*, Boston, MA, May 1994.

24. T. Wahl, S. Wirag, and K. Rothermel. Tiempo : Temporal modeling and authoring of interactive multimedia. In *Proc. 1995 International Conference on Multimedia Computing and System*, Washington, DC, May 1995.

25. R. Weiss, A. Duda, and D.K. Gifford. Composition and search with a Video Algebra. *IEEE Multimedia*, 2(1), 1995.

26. R. Willrich et al. Hypermedia document design using the htspn model. In *Proc. Third Int. Conference on Multimedia Modeling*, Toulouse, France, 1996.

An End to End Price-Based QoS Control Component Using Reflective Java

Jérôme Tassel, {jtassel@jungle.bt.co.uk}
Bob Briscoe, {rbriscoe@jungle.bt.co.uk}
Alan Smith, {asmith@jungle.bt.co.uk}
BT Advanced Research & Technology, UK

Abstract

The main objective of the model we describe in this paper is to allow easy, flexible addition of quality of service (QoS) control to Java Internet applications. In this work the QoS is expressed in terms of network and host resources, the network QoS being controlled with RSVP. Flexibility is provided by a prototype product from the ANSA research consortium; Reflective Java which uses the Meta Object Protocol (MOP) to separate functional requirements (what the application does) from non-functional requirements (how it does it). This protocol permits the design and implementation of a generic QoS control element which can be added to an application for which QoS control is required. Alternatively, an existing application with rudimentary QoS control can be modified to use a set of QoS control classes designed by a specialist intended to reconcile competition for QoS between applications. The QoS control element we have designed also has scope for QoS adaptation, moving decisions on the introduction of QoS control from build-time to run-time when best-effort degrades below a useful point. Charging is also considered in this work.

Acknowledgements

This work was funded by BT as part of an industrial placement agreement for Jérôme Tassel from the MSc in Distributed Systems course at the University of Kent at Canterbury (UKC). Bob Briscoe and Alan Smith are researchers in the Distributed Systems Group in BT's research labs. The authors would like to thank Steve Rudkin, Peter Bagnall and Andrew Grace at BT, Zhixue Wu working on the ANSA project and Andy King from UKC for reviewing earlier versions of this paper and their valuable advice.

1.   Introduction

In this paper, we describe the design of a flexible, easy-to-add adaptive quality of service (QoS) architecture for multimedia, Java-based [19,20], Internet applications. Control is provided for QoS properties that we identified as crucial in specifying QoS for real-time applications. This set of properties can be divided into two groups :

- User requirements (prioritisation, quality perception, budget)
- Mechanisms (Network QoS, RSVP in this case, and Host QoS)

Flexibility and adaptability is provided by a prototype product from the ANSA research consortium; Reflective Java which uses a Meta Object Protocol [12, 13, 14, 15] (MOP) to separate functional requirements (what the application does) from non-functional requirements (how it does it) [15]. The generic QoS control element we have defined can be added to an application for which QoS control was not originally thought of or to replace a deficient QoS control or to provide QoS control as a result of adaptation to network conditions [28]. Applications which require such a control are emerging Internet collaborative tools with a multimedia interface using audio and video streaming facilities. The QoS control element we have built also has scope for

QoS adaptation and charging. We believe this model provides a flexible way to control the usage of the resources available to the user down to a fine level of granularity. A QoS control interface is available to the user giving him complete flexibility over the sharing of his resources among his application sessions, media sessions and streams. This interface might exist as an operating system component separately from any applications that might use it.

The first half of the document describes RSVP (Internet ReSerVation Protocol) and Reflective Java, focusing on the points of interest for this project. The second describes in depth the architecture we have designed and some of the implementation results and experiences we have had so far.

2. QoS Control on the Internet with RSVP

RSVP is used in this work as the network QoS control mechanism. The design allows for other mechanisms to be used instead, or as well. Effectively RSVP allows control of the quality of service of data streams over the Internet [2,4]. Internet applications are changing from simple remote procedure call (RPC) type text based point to point applications to real time, multi-user, multimedia applications [1]. The original design of the TCP/IP suite only provided support for a best effort delivery scheme, ideal for applications such as ftp, World-Wide-Web, Telnet, and e-mail but is very deficient for real-time delivery of data. The efforts of the Internet Engineering Task Force (IETF) are now focused on developing a new Internet architecture which can provide Integrated Services on the Internet (the ISA). This will allow a wide range of QoS to co-exist, some with hard or soft real-time delivery constraints and some with more elastic timing constraints. The main weakness of the IP protocol for guaranteed delivery on the Internet is the variable latency of packet processing time within the routers along the data path, due to queuing delays which increase the jitter of the data stream, thus jeopardising real time delivery of data. The objective of RSVP, as part of the ISA, is to provide some control over the routing queuing delays and therefore the QoS of data streams as they pass through the routers.

RSVP attempts to reserve bandwidth for real time streams across the routers by setting packet priorities. In the case of guaranteed delivery of streams, RSVP allows provision of a guaranteed set of resources for a data stream based on an IP address (which can correspond to a unicast or multicast address) and a port number. We now describe the major features of the RSVP protocol which are of interest for our work.

2.1 Receiver Initiated Reservation and Message Processing

An Internet communication application can be divided into two parts, the sending and the receiving part(s). The sending application acts as a source of data for the receiving application(s); there might be multiple receiving applications in a multicast communication environment. Multiple senders can be treated as separate sources and collected or synchronised independently if necessary. In order to accommodate heterogeneous receivers on the Internet, receiver initiated reservation has been chosen in RSVP. The receiver chooses the reservation it wishes to make for a given source data stream. It uses information disseminated by the sending application about the data it is producing to decide what reservation is required. Admission and policy control is the subject of current research work [5] to provide administrative control of bandwidth sharing and to make sure that streams keep to their agreed traffic properties.

Two types of messages are defined in RSVP, namely 'PATH' and 'RESV' messages. PATH messages are used by the sending applications to disseminate information about their data streams, and RESV messages are used to establish the reservations by the receivers that get the PATH messages.

2.2   QoS Attributes

RSVP does not know about QoS attributes but just conveys them transparently [5]. There are two main types of service available for RSVP: a guaranteed service and a controlled load service In our work we use the guaranteed service. The reservation attributes in the PATH messages define the sending application requirements in terms of expected bandwidth usage. The reservation attributes in the RESV messages represent the required bandwidth reservation for the receiving applications. The attributes used for the control of the bandwidth are the generated traffic rate and peak rate, the token bucket rate, minimum policed unit (defining a minimum size for packets to be policed) and the maximum packet size (for the underlying network).

2.3   The RSVP API (RAPI) [3]

An application programming interface (API) is available to interact with the RSVP daemon on hosts which implement RSVP. Operations are available to *join* a session (an RSVP session), register as a *sender* and send PATH messages, *reserve* bandwidth, modify the reserved bandwidth and *release* a reservation.

Another important aspect of RSVP is that of *event messages*. These messages are asynchronous (but synchronous results are also provided by the API calls, which provides some basic error checking such as parameter checking). Errors and modifications might occur along the data path and messages are sent back through the RAPI in order to notify the applications using RSVP for them to take appropriate actions. Messages might indicate an arrival of PATH messages, RESV messages, path errors, reservation confirmation or not. These upcalls are associated with an application level method which is called on receipt of such events.

2.4   QoS Adaptation in RSVP

The PATH and RESV messages we describe above are sent periodically (the RSVP daemon handles this) so that if the properties of the sending application stream are modified or the route through the network changes the receiver can then adapt its reservation. An RAPI call is available to modify the attributes of the reservation.

Aside from RSVP, Reflective Java is the other key element of an architecture which needs to be understood and is therefore described in the next section.

3.   Reflective Java

Reflective Java is a prototype product from the ANSA research consortium [16, 17] providing the facility to separate the functional aspects of an application from the non functional ones [11], this is done using the Meta Object Protocol (MOP) [12, 13, 14, 15]. This separates the roles of application developers, one focuses on what the application does and another on how it does it. The term MOP is unfortunate - it is not a communications protocol, more a design pattern allowing this separation. Examples of non-functional requirements are the well know transparency requirements of the distributed systems world: replication, concurrency, failure transparency, some of which have also been worked on in a very similar development called MetaJava [18]. This scheme means that some requirements can be added late in the life cycle of an application, even if they had not been thought about originally.

The following diagram illustrates how Reflective Java implements the Meta Object Protocol at run-time:

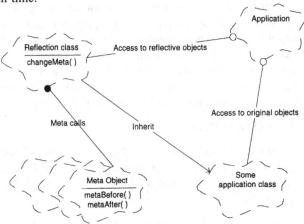

Fig. 1, The Reflective Java run-time model

The *meta-objects* implement some non functional requirements and are independent from any application which uses them so that they can be used by many applications. The meta-object developer can focus on designing meta-objects providing some functionality and those objects are later bound to application code. The meta-object defines two methods: *metaBefore* and *metaAfter* which implement some behaviour to be executed before and after the normal invocation of an application method. For example a locking meta-object providing concurrency control would provide some code to set a lock in the *metaBefore* method and unlock it in the *metaAfter* method.

The *Reflection class* is used to bind applications to meta-objects (bind the meta methods to the application class methods). A simple example is a locking meta-object and an account application class which uses this meta-object to provide concurrency control on the accounts. So for example a deposit method would be bound to the locking service provided by the meta-object. The original invocation of the deposit method would then become :

| Application call deposit() | → | Reflective class calls metaBefore | → | Reflective class calls original deposit method | → | Reflective class calls metaAfter |

When the service provided by the meta-object is needed by the application class, the reflection class is used in the application code instead of the application class it inherits from.

As shown on the above diagram another important feature of Reflective Java is that the meta-object can be *dynamically* replaced by another one, implementing a new version of a service or a new service being more adapted to a new environment. This overcomes the drawback of having to manually update the code of the application to be able to make use of the reflection class as there is no need to further modify the application when the services from another meta-object are required.

In the build process, it is the reflective pre-processor which creates the reflective class from a binding specification file, which dictates which application method class should use the services of the meta-object. The following diagram illustrates the build process of Reflective Java.

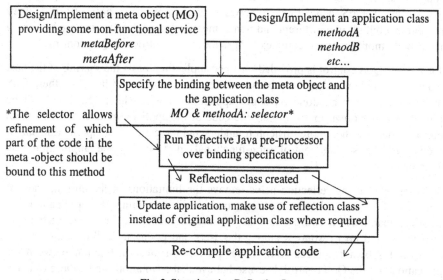

Fig. 2, Steps in using Reflective Java

This implementation of the reflection technique in Java is made possible by the dynamic class loading feature of Java. The possibility to download Java classes over the network make this scheme very flexible and place Reflective Java ahead of some other implementations of Reflection described in [12,13,14,15].

In the work presented in this paper we use the capabilities of Reflective Java to separate the QoS control of Internet applications from the mechanisms of those applications; sending data.

4. QoS Architecture

The aim of our architecture is to be able to provide a generic (adaptive) QoS control element, which allows the best use, in terms of user-perceived value, of the host and network resources [26]. However, it is of utmost importance that this element can be added to existing Internet communication applications written in Java by an independent developer. This implies it is not necessary to know the detailed internal mechanisms of the application and the integration could be reduced to a mechanical task in the future. Source code access is required, however.

Currently, for applications without any QoS control [29], as the network or host becomes congested the quality of the communication decreases due to the unreliable nature of the Internet best-effort data transfer protocols and the lack of resource control in most current operating systems. Today's Internet communication applications, written in Java or any other language, provide functionality both for sending the data and controlling the quality of the data sent. Typical examples are the vic and vat video/audio conference tools [6, 7] which provide a control panel for the properties of the multimedia streams. These applications do not make use of any network or host resource reservation protocols, although a recent prototype of vic now supports RSVP. Some others such as Vosaic or QuickTime Conferencing provide some form of QoS adaptation to network congestion in a best effort delivery environment but all in a different and non compatible manner and some are becoming increasingly monolithic and complex [31] as they try to introduce QoS control.

Moreover as QoS control is provided on a per-application basis the sharing of the host resources available to the user between users will not be optimal. Further, QoS adaptation can only be done on a per application basis, which again will not be optimal. Another downside of mixing the code of the application task and the stream control is that it is then more difficult to understand the code of the application and to modify just the stream control (non-functional) part or just the application task (functional) part.

Our component using reflection helps remove the limitations of the current ways of integrating QoS control with real-time multimedia streaming [8] applications by providing guaranteed network resources over the Internet as well as sharing efficiently the network and host resources among the users and giving a flexible control to the user over his own resources. Our model provides a clear distinction between the application and the QoS control. We also believe that adaptation can be more efficient as more host-wide information is available to the adaptation mechanism.

The diagram on the following page is a representation of the model we have designed. There are three major elements in this model:

1. The original application (which models typical Java Internet applications)
2. The QoS control architecture (which includes components to control host and network resources in order to share them between application streams)
3. The reflective architecture (which binds the two previous parts together)

In the model, different components control or give quotes for a unique type of resources (host or network). The QoS manager acts as a conductor of all of them. Resource control and quoting have been separated to permit them to evolve independently. The model we designed is for receivers but the same techniques could be adapted to senders. However we have only implemented the former. The following section describes the design choices for the model we created.

4.1   Original Application (1)

The aim of this part of the model is to represent a typical Internet application that uses some communication class to interact with the network. The original application classes altogether provide both the functional and the non-functional requirements of the application. The original application [29] (that is, without our QoS control) would only use the communication class and not the reflection class. The reflection class is the reflective element that allows us to include QoS control in the application with minimal modifications to the code.

At the application level, an important issue related to the reflective architecture was to decide which class in the application to make reflective. We were faced with two options:

- to make the communication class used by the application reflective
- to make a more high level application class reflective.

In order to make our solution reusable and portable we decided to take the first solution In the case of Java this meant making the Socket class reflective. This means that when the application code is to be modified there is no requirement to know the structure of the program but just to know where communication objects are instantiated and to replace the original communication class by the new reflective class we have designed. It also means that any application using the Java Socket class can re-use the work we have produced (in the other case a new binding specification would have been required for each new application).

This, of course, means that the application needs to be modified, which is even more constraining as it is done *manually* (however, a modified class loader could be used to avoid this). The main issue during this process is to decide *which communication objects require QoS control* as not all streams need timeliness; some are signalling streams for which a best effort delivery scheme is very well suited. We identified the following options:

- *Make all communication objects (Sockets) reflective but test the port number within the meta object class and set reservation only on some well known ports.* A port number is an obvious selection criterion as it is the only element of information guaranteed to be available when any general communication channel is created. This information can be passed to the meta-object which will then decide on the path to follow; interacting with the QoS manager or not. This has got some run time cost as all communication objects would be reflective while only some would really need to be. Also some applications use random ports in which case this scheme could not be used. However , the main advantage of this method is that it requires only a few easy changes to the application code.

- *Make all communication objects reflective but include heuristic logic in the meta-object to decide at run-time which calls do or do not require real-time control.* This has the performance cost drawback of the previous solution but avoids the need for port number conventions. It would be the only solution if the need for

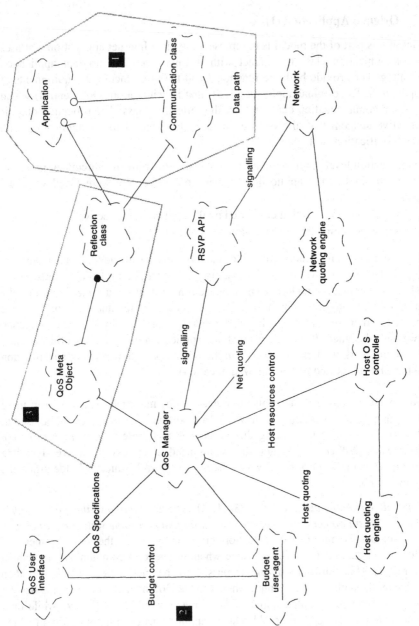

Fig. 3 , Reflective model for (adaptive) QoS control

real-time control for each class varied depending on run-time conditions. Another drawback is that it is not clear to us how one could define the heuristics.

- *Make all communication objects reflective as the channels which do not require real-time control will not receive PATH messages and therefore reservations will never be established.* This again has the same drawbacks as the previous solution and worse, RSVP sessions will be created which will never be used, but the solution again requires no conventions concerning port numbers.

- *Give control to the user to state whether or not a stream is to be timely.* This would present the user with a stream control panel. This of course is a too finely grained solution as the user should not be aware of the stream notion and will probably not know when a stream is to be a data stream which needs to be timely or a signalling stream which does not require real time control.

- *Go manually through the code and decide according to the context if the stream requires QoS control or not.* This means that the code of the application must be clear enough for the system integrator to decide if the data transmitted on a stream is real-time or not. In our case this is the solution we have chosen due the prototype nature of the target applications we use. However, this approach would not be appropriate if the need for real-time control varied with run-time conditions.

4.2   The Reflective Architecture (3)

The reflective part of the model allows us to link together the application services and the QoS control architecture we have designed. Two classes make up this part of the model, the *QoS meta-object* implements the QoS control which in our case is delegated further to the QoS manager. As a result the role of the meta-object is only to call the QoS manager. The call to the QoS manager is in the metaAfter method so that the creation of the socket is not held up, the metaBefore method is used to release reservations on destruction of a socket. Furthermore making a call to the QoS manager (and not directly to the RSVP API) does not tie down our Reflective model to the use of RSVP.

Notice that thanks to the dynamic facilities of Reflective Java the meta-object could be modified at run-time. For example, when the environment changes (from a fixed to a mobile network, or from a network using RSVP for bandwidth reservation to another using another reservation protocol [9, 10, 25]) a method is provided to switch to a new meta-object in response to such arbitrary events. This could be used to call a different QoS manager more appropriate for the new environment or having some extra-functionality. The QoS meta-object is generic, it is not designed with any host applications in mind. We have added a call to the QoS manager (described in the following section) in the metaBefore method and in the metaAfter call. The call in the metaBefore method is to set-up the required QoS for the newly created stream (the key used to identify each stream is "user name + application name + stream IP address + port number") and the call in the metaAfter call is to release reservations.

The second class is the *reflection class,* which is a refined version of the original application communication class and binds together the meta-object methods to the appropriate methods of the original application class. The reflection class is specific to the application class we decided to make reflective; the Socket class. It was created originally by the Reflective Java pre-processor but we had to make manual changes to it due to some limitations of ANSA's Reflective Java prototype (no support for a Reflective constructor or make classes from the Java APIs reflective).

4.3   The QoS Control Architecture (2)

The aim of the QoS control architecture is to manage resources on integrated service networks (focusing on Internet and RSVP). The resources our architecture controls are network resources (in terms of bandwidth), host resources (in terms of memory, CPU usage and hard disk space) and user resources (budget). Fig 3 illustrates the different objects of our architecture. The model we have designed is host oriented; the role of the QoS manager is to co-ordinate the services of the other QoS related objects in order to manage resources on behalf of a host. Therefore on a network using our model, each host would hold a copy of the full model described in Fig 3. Each user on the host could be running multiple applications using many streams, some of which would require QoS control. There would therefore be many instantiations of the QoS meta-object class and reflection class, one for each of the streams which required QoS control. But each user has access to only one QoS user interface, which integrates control over all the streams he owns. Also there is a single user budget agent per user.

Implementation issues regarding communications between objects running on different virtual machines (the user applications and the user QoS control interface for example) have been solved by using a product internal to BT: the JavaShell. With this, multiple applications can run on the same virtual machine and communicate with each other (in our case the QoS control objects, which are unique for a host, are static and can be accessed by all other objects). This avoids the need to differentiate between real host resources and those allocated to each virtual machine, a problem likely to disappear as the virtual machine becomes integrated with the operating system [21].

The architecture is not exhaustive in terms of how resources can be managed but objects could easily be added to control resources differently (e.g. administrative QoS management, or adding billing facilities). The next paragraphs discuss the different elements of this architecture.

Managing network resources

In order to control network resources for the application streams, the QoS requirements of those streams must be defined. As our QoS model can be added to an application which does not deal with QoS control (or very little), those requirements must be emanating from another source. As we use RSVP to manage the network QoS, we use the information provided in the TSpec element of the PATH messages to decide on what QoS properties a stream might require. Those requirements are then balanced by the QoS manager according to the user choice of priorities between streams (see below) and the resources available (on the network and the host) [22]. Juggling of user priorities is done in financial terms. The costs of network resources are therefore needed; these are provided by the network quoting object. The assumption is that, in the future, all QoS will have a price in order to put a brake on profligate users.

It is the QoS manager which issues demands for bandwidth reservation via the RSVP API; it holds the set of users, their applications, their streams and the network QoS for these streams. Notice that, because our implementation is in the Java language we had to implement a RSVP API in Java using Java native (JNI) calls [24].

Streams that are controlled can remain in a best effort mode until they require reservation, as the network or the host becomes congested. Another approach would be to provide them with a set of default resources as defined by the TSpec element of the RSVP PATH messages. The user by changing relative priorities between streams will then trigger modification of reservations.

A QoS monitoring object could signal the QoS manager when a stream is not performing well enough and then the QoS manager would provide some reservation for this stream. In our current implementation, the user can trigger reservation by using the QoS control interface we have designed, but integration of our automatic QoS monitor [23] is outside the scope of this work.

Also we decided to implement an application API for some of the QoS manager services so that QoS controlled directly from an application can also be taken into consideration. The QoS manager user interface shown in Fig 5 will be enhanced to make the distinction between application and user control of stream priorities.

Managing host resources

The role of the host controller object is to provide some admission control facilities regarding the usage of the host resources by the application streams on behalf of the users. We designed and implemented a host resources control object, which holds the amount of available resources on the host and registers their use. We also implemented a host quoting engine which returns a price from a specification of host resources. The host QoS requirements for an application stream is deduced from the required network QoS for the same stream. This might seem simplistic but it prevents overloading of the host. A more complex version of this object could be implemented but the services required would still remain the same as we have implemented. The QoS manager interacts with the host controller to set and release reservations and the user budget agent uses the services of the host quoting engine to authorise reservations.

Managing user priorities

Fig. 4, The user QoS control interface
Note that the identification of streams by 'address:port' is merely a prototyping convenience -
textual stream ids such as 'audio' or 'video' could have been generated relatively easily

The figure above shows the user QoS control panel we have implemented. This interface is available for each user to define the priorities between applications and streams within applications. The user can select an application in order to get the list of streams created by the application (the Sockets instantiated from the reflective class of the reflective part of the model). Then he can define which stream should have more priority over the host and network resources by positioning the scroll bars accordingly. These priorities have an impact on the distribution of the user budget over the current streams he is using. For users not wanting to have such a finely grained level of control on the streams, we have added a user preference which removes the list of streams from the control panel and only provides control over the list of applications. The resources are then distributed over the applications in proportion to the resources the PATH messages said were needed. Development of the GUI to express distinction between application requirements on priorities and user override is ongoing.

These priorities are then sent to the QoS manager which increases or decreases the resources reserved for the related streams in conjunction with the quoting engines, the resource controllers and the user budget agent (network and host). Streams being defined with a null priority are left in a best effort communication mode. This control could be part of a host control panel (similar to the one in MS Windows) and the QoS architecture could be part of the host operating system to provide a QoS management service to the user.

Managing charging issues

We have introduced a user budget agent in our design, which controls the budget of a user according to some rules predefined by him. The rules include the time constraints on reservation establishment and restrictions on maximum spending per stream and per application . When a new set of resources needs to be reserved on behalf of a stream, the QoS manager decides which attributes to reserve and mediates between the quoting engines and the user budget agent. The QoS manager requests a quote for the amount of resources it wants to reserve and sends this quote to the user budget agent which then can authorise or refuse it, according to the user spending rules.

It was a deliberate decision to normalise all QoS comparison into units of relative financial cost rate. This is the most convenient common unit both for the programmer and in terms of user understanding.

5. Conclusions and Further Work

One of the main achievements of our work is a component with a clear separation between the transmission mechanisms of an Internet application and the policies and mechanisms to control its QoS. This makes it very flexible in terms of later upgrades, configurability and adaptability. An Internet programmer can focus on the mechanisms of the application and a QoS specialist can focus on providing a QoS control element, thus increasing productivity. This separation also means that existing applications as well as forthcoming applications and the QoS control model we designed can easily be integrated. Re-usability is also an important achievement in our model as meta-objects are application independent. Our model also provides scope for adaptation to a changing environment. We only designed it for Internet (RSVP) but another one could be designed for a different environment (e.g. ATM, mobile networks) We have provided the run-time switching mechanism for this.

In our model, the presentation of QoS control to the user proved to be difficult. Our work has proved to us that relative priorities between cost rates are the most useful concept in this respect. Reflective Java proved to be relatively immature. Readers interested in a constructive critique of Reflective Java for QoS control can refer to [30] where a more in depth description of this work is also available.

The implementation of the model is still continuing as at this first stage we concentrated on getting the overall architecture working. We aim to provide better control in a multicast communication environment, provide more event-based reactions by the QoS manager and finally we would like to develop a reusable RSVP API in Java because the one we implemented for this work is very specific to our QoS manager. Another area of possible extension is in the run-time heuristic decision over whether to make a socket reflective (and thus give it timeliness control) and finally the QoS manager could be further separated to avoid over-centralisation.

6. References

[1] RFC 1633, Integrated Services in the Internet architecture: an overview. R. Braden, D.Clark, S.Shenker. June 1994. (33 pages) http://www.isi.edu/div7/rsvp/ietfpub.html

[2] RSVP: A new resource reservation protocol. Lixia Zhang, Stephen Deering, Deborah Estrin, Scott Shenker and Daniel Zappala. IEEE Network. September 1993. (9 pages) http://www.isi.edu/div7/rsvp/ietfpub.html

[3] RAPI -- RSVP Application Programming interface -- version 5. IETF RSVP Working group. Draft RFC, Expires November 1997. R. Braden, D. Hoffman. http://www.isi.edu/div7/rsvp/ietfpub.html

[4] Resource ReSerVation Protocol (RSVP) - Version 1 functional specifications. Internet draft. R. Braden, L.Zhang, S.Berson, S. Herzog, S.Janin. November 1996. (82 pages) http://www.isi.edu/div7/rsvp/ietfpub.html

[5] The use of RSVP with IETF Integrated Services. J. Wroclawski MIT LCS. Internet draft. Expired 3/97 http://www.isi.edu/div7/rsvp/ietfpub.html

[6] vic: a flexible framework for packet video, McCanne S and Jacobsen V, Proc of ACM Multimedia '95, Nov 1995.

[7] First IETF Internet Audiocast, Casner S and Deering S, Computer Communications Review 22, July 1992

[8] Specifying QoS Multimedia Communications within Distributed Programming Environments. Daniel G. Waddington, Geoff Coulson and David Hutchison. Computing Department Lancaster university. Proceedings of the 3rd International COST237 Workshop. Volume 1185, pp. 10-4-130. November 1996(20 pages)

[9] Supporting adaptive services in a heterogeneous mobile environment. Nigel Davies, Gordon S. Blair, Keith Cheverst and Adrian Friday. Distributed Multimedia Research Group at Lancaster University, ADAPT project. In proceedings of the 1st workshop on mobile computing systems and applications, Santa Cruz, CA, December 8-9, 1994. (5 pages)

[10] Quality of Service control for adaptive distributed multimedia applications using Esterel. Andrew Grace, Alan Smith. BT Laboratories. (8 pages) Second International Workshop on High Performance Protocol Architectures, Sydney, 11-12 December 1995

[11] A QoS Configuration System for Distributed Applications. Alan Smith, Andrew Grace, BT Laboratories. 5th IFIP International Workshop on Quality of Service (IWQOS'97) (4 pages) http://comet.ctr.columbia.edu/iwqos97/

[12] Concepts and experiments in computational reflection. Pattie Maes. AI-Lab, Vrije Universiteit Brussel. Object-Oriented Programming, Systems, Languages and Applications (OOPSLA) Proceedings, 4-8 October 1987. (8 pages)

[13] Computational Reflection in Class based Object Oriented Languages. Jacques Ferber, LAFORIA. Object-Oriented Programming, Systems, Languages and Applications (OOPSLA) Proceedings, 1-6 October1989. (8 pages)

[14] Workshop: Reflection and Metalevel Architectures in Object-Oriented Programming. Organiser: Mamdouh H. Ibrahim. Object-Oriented Programming, Systems, Languages and Applications (OOPSLA/ECOOP) Report, 21-25 October 1990. (5 pages)

[15] Atomic Data Types. T.J. Stroud and Z. Wu. Department of Computing Science, University of Newcastle upon Tyne, UK.

[16] Design of Reflective Java. ANSA draft, Zhixue Wue and Scarlet Schwiderski, 23rd December 1996. APM.1818.00.06 (22 pages) Restricted to ANSA sponsors http://www.ansa.co.uk/Research/ReflectiveJava.htm

[17] Design and Implementation of a persistence service for Java. ANSA, Scarlet Schwiderski, 27th, January 1997. APM.1940.02 Restricted to ANSA sponsors http://www.ansa.co.uk/Research/ReflectiveJava.htm

[18] MetaJava: An efficient run-time meta architecture for Java. Jurgen Kleinoder, Michael Gölm. Friedrich-Alexander-University, Erlangen-Nurnberg, Computer Science Department, Operating Systems-IMMD IV, Germany, June 1996. TR-I4-96-03. http://www4.informatik.uni-erlangen.de/Projects/PM/Java/

[19] The Java language Environment, A white paper. James Gosling, Henry McGilton. Sun Microsystems Inc. October 1995. http://java.sun.com/docs/index.html

[20] The Java Platform, A white paper. Douglas Kramer. Sun Microsystems Inc. May 1996. http://java.sun.com/docs/index.html

[21] Secure computing with Java: Now and future. Sun Microsystems Inc, 4th June 1997. http://java.sun.com/docs/index.html

[22] QoS Mapping home page and set of supporting slides. Daniel G Waddington, BT-URI at Lancasteruniversity, End-system QoS in Multi-service Networks project. 6 December 1995 http://www.comp.lancs.ac.uk/computing/users/dan/uri/mapper/index.html

[23] Support Components for Quality of Service in Distributed Environments: Monitoring Service, D.A.Reed and K.J.Turner, Stirling University, UK, 5th IFIP International Workshop on Quality of Service (IWQOS'97) (4 pages) http://comet.ctr.columbia.edu/iwqos97/

[24] The Java native programming interface. Sun Microsystems Inc http://java.sun.com/docs/books/tutorial/native1. 1/implementing/index.html

[25] Campbell, A.T., "QOS-Aware Middleware for Mobile Multimedia Networking", Multimedia Tools and Applications, Special Issue on Multimedia Information Systems, 1997, (to appear) http://comet.ctr.columbia.edu/~campbell/andre w/publications/publications.html

[26] Aurrecoechea, C., Campbell, A.T. and L. Hauw, "A Survey of QoS Architectures", Multimedia Systems Journal , Special Issue on QoS Architecture, 1997, (to appear) http://comet.ctr.columbia.edu/~campbell/andre w/publications/publications.html

[28] Campbell, A.T., Coulson G., and D. Hutchison, "Supporting Adaptive Flows in Quality of Service Architecture", Multimedia Systems Journal, Special Issue on QoS Architecture, 1997, (to appear) http://comet.ctr.columbia.edu/~campbell/andre w/publications/publications.html

[29] Web2Talk, Internet telephony application. Roger Klein, Carsten Schulz-Key, Stephane Chatre http://www.CS.ORST.EDU/~kleinro/Web2Talk

[30] QoS Adaptation using Reflective Java Jérôme Tassel. M.Sc project dissertation, University of Kent at Canterbury UK. September 1997. To appear.

[31] Implementation of stream module in a Distributed Computing Environment. Rainer Aschwanden, BT Labs. September 1996.

SECCO - Support Environment for Electronic Commerce

V. Tschammer, M. Mendes[1], V. Ouzounis, M. Tschichholz

GMD FOKUS Berlin

Kaiserin-Augusta-Allee 31

D 10589 Berlin

Abstract

The paper describes SECCO, a support environment for electronic commerce applications. The environment is based on distributed object technology and is composed out of existing building blocks. Most of these building blocks have been developed according to OMG CORBA and TINA-C specifications and are being used in an heterogeneous environment. The paper presents a brief overview of electronic commerce, discusses related activities, and then presents the SECCO architecture and describes the available components. An application example finally illustrates the use of SECCO in a collaborative scenario.

Keywords: Electronic Commerce, Distributed Object Technologies, Support Environments

1 Objectives of SECCO

SECCO, the Support Environment for Electronic Commerce is being developed by GMD FOKUS within the framework of its "Y++" Research and Development Programme. SECCO will provide an interoperable infrastructure and architecture supporting the development and operation of application components within Electronic Commerce scenarios.

SECCO specifically - but not exclusively - addresses those applications which are based on the distributed object-oriented computing paradigm and/or mobile agent technology. According to the state of the art, the prime technologies to be considered in these areas currently are OMG CORBA, Java, and OMG Mobile Agent Facility (MAF).

Based on these technologies we concentrate on issues and developments which are in line with the approaches considered by the OMG Electronic Commerce Domain Task Force (ECDTF), the ACTS OSM project, and the CommerceNet Initiatives.

Interoperability with major Electronic Commerce (EC) platforms (IBM CommercePoint, Netscape ONE, Oracle NCA, Sun/Javasoft JECF, and Microsoft Internet Commerce Framework) has been discussed in (1) and is to be achieved based on CORBA IIOP, Java, a common distributed object model, and protocol negotiation, gateways, and mediators.

We also assume that within SECCO the EC applications and support facilities will be closely related to the Business Object (BO) concept (2). Therefore, we expect that

---

[1] On leave from the Pontifícia Universidade Católica de Campinas, SP, Brasil.

SECCO will be based on BO principles and will be realised by means of Business Object Facilities (BOFs) and Common Business Objects (CBOs) – if available.[2]

2   Introduction to Electronic Commerce

A possible definition of Electronic Commerce is 'any form of business transaction in which parties interact electronically rather than by physical exchanges or direct physical contact'(3).

Included in such business transactions are transactions within a single company, electronic trading between a supplier of goods and its customers, and all forms of open service markets and information provision. Such business transactions are much more decentralised in organisation than the traditional ones, i.e. not only enterprises but also departments, groups, and individuals are involved in doing business via electronic communication facilities. Electronic commerce, thus, is lowering the barriers between enterprises, suppliers, and customers, making traditional business being re-designed into co-operative activities which are jointly owned and operated by different parties.

Electronic commerce is characterised by diversity. A wide range of business operations and transactions have to be included. We can distinguish between:

- the business processes and transactions itself,
- the communication and information technologies used,
- the federation and retailing issues imposed by joint business processes, and
- the legal and regulatory framework.

SECCO is mostly concerned with the three former issues.

Business processes and transactions. Relevant business processes and transactions include:

- the establishment of the initial contact between the parties involved, and the exchange of information,
- the pre- and post-sales support, such as details of available products and services and the technical guidance on product use,
- the negotiation and the making of contracts,
- the subscription to services, the control of service sessions, accounting and billing,
- sales and electronic payment using EDI, credit cards, electronic cash or checks,
- distribution of products that can be delivered electronically as well as the management and tracking of physical products,

---

[2] BOF proposals are still in a preliminary state and CBOs are still to be defined.

Communication and information technologies. In these processes and transactions a wide range of communication and information services is involved. Current technologies are email, fax, EDI, and EFT. Directories and document interchange are also mentioned, as well as security and multimedia. These technologies are to be enhanced and integrated by the platforms envisaged, and SECCO is going to provide a set of EC support facilities and services which are to support the processes listed above.

Federation and retailing. Federation and retailing are further issues within the EC context. They are imposed by the fact that the traditional business is being replaced by co-operative processes which are jointly owned and managed by different parties. Relevant issues include:

- virtual enterprises, i.e. the grouping and co-operation of individual enterprises so that they can offer products or services that would be beyond their individual capacities,

- shared business processes that are jointly owned and operated by an enterprise and its trading partners.

3   Relevant Approaches

3.1   The OMG Domain Task Force on Electronic Commerce

Several current activities at the OMG have strong connections with Electronic Commerce. The most important are Business Object Domain Task Force, Financial Domain Task Force, Common Facilities Task Force, Object Analysis and Design Task Force, Electronic Commerce Domain Task Force (EC-DTF).

Recently, the EC-DTF published an Electronic Commerce Reference Model and Component Architecture (4) which has been prepared as a work-in-progress. It identifies and describes the components of the reference model and establishes the correlation with the OMG Object Management Architecture, the Object Services, and the Platform and Domain Facilities.

The architecture is composed out of three principal groups, namely (a) low level electronic commerce services including payment, profile and selection services; (b) commerce facilities supporting contract, service management and related desktop facilities; and finally, (c) market infrastructure facilities covering catalogue, brokerage and agency facility.

3.2   Principal Modules and Facilities of the EC-DTF Component Architecture

Profile (Interchange) Facility. A profile is a data object that can be considered as a portable, persistent container for arbitrary data objects. It comprises standardised information that is interpreted by service market infrastructure components, such as object browsers, catalogue services, brokers or traders. The profile facility supports the building and modification of profiles, as well as their exchange between EC participants.

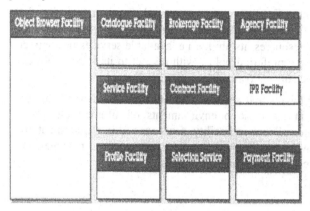

Electronic Payment Facility. Under the ECDTF RFP-1, the OMG is soliciting technology which can provide a payment facility that will support and allow for implementation of a variety of electronic payment protocols in a CORBA Distributed Object Computing environment.

Selection/ Negotiation Facility. The selection/ negotiation facility is to support the selection and configuration of supporting facilities across the domains involved in an electronic commerce transaction. It will provide a gateway between the macro-policy agreed for the commercial transaction, and the local policies of the commerce participants.

IPR Facility. IPR Facilities provide the support for the management and administration of Intellectual Property Rights including copyright and ownership.

Contract Facility. Contracts encompass both the static contractual perspective of a real-world contract (as a data model), and the dynamic perspective in the context of a contract as an instrument supporting the execution of a set of business processes (presented as commercial services) under the context of a common and potentially dynamically changing policy. The contract facility supports contract re-use and contract encapsulation. It provides an interface that is capable of hiding the details of the contract structure, signatures, and related terms and conditions.

Service Management Facility. The service management facility provides an object framework which supports the separation of declared interfaces between the consumer, the provider, and any third-party involved in an electronic commerce transaction. Functions include service life-cycle support, service policy management, service presentation and inspection, service subscription and accounting., as well as service invocation and session support.

Catalogue Facility. The catalogue is a structured object that can be inspected, browsed and transferred through the network. It provides an interface that allows to re-order the catalogue entries in different views, to extend the catalogue by additional entries, and to modify and existing entries. A major application for a catalogue is to contain service and contract information.

Brokerage Facility. Electronic commerce is the ability to perform exchanges of goods, services, content, assets, and money between two or more participants using electronic tools and techniques. The brokerage facility supports EC participants so that they can communicate, navigate, gather, search, filter, deliver, and route information, and exchange contracts, services and other commerce-related objects.

Agency Facility. The Agency Facility supports general requirements for the standardisation of a point of presence in a market. It administers and provides information about the agency resources, its ability, i.e. available services or contract templates, and its policy, i.e. the preferred policy with respect to its local facilities, e.g. the preferred payment mechanism, available security level, etc.

Object Browser Facility. Products and services in electronic commerce must be portable and inter-operable in the desktop environments of different electronic commerce systems. The Object Browser Facility deals with this requirement by providing the desktop framework for the presentation, inspection, and management of components such as services, contracts, and certificates.

3.3   The OSM Project

The ACTS OSM project (5) develops an open architecture for electronic markets and a supporting implementation of that architecture. The architecture and infrastructure are to be validated through product trials.

The prime objective of the OSM project is to accelerate commercial opportunities through market growth around compliant electronic commerce tools and infrastructure facilities. This objective is to be achieved through adoption of the OMG EC-DTF standards.

The OSM reference architecture is composed of four principal layers covering general electronic commerce facilities at their base, migrating up to facilities supporting collaboration between participants, through a commerce layer supporting facilities such as contracts and compound services, and finally, a market layer which supports agency facilities.

The demonstrator implementation includes an electronic market scenario with service offers, service mediation mechanisms, and service access facilities. It currently comprises software components for profile management, profile editing, persistent asynchronous communication, online catalogue for service offers, service trading, generic client, automatic matching and selection of support services.

The results of the demonstrator implementation will be integrated into the OSM infrastructure and will serve as a framework that allows for further application development such as a mobile agent system and an application management framework.

3.4   The Eco System: An Internet Commerce Architecture

The Eco system (6) is an industry effort to build a framework of frameworks, involving both EC users and vendors. It has been initiated by the CommerceNet consortium which is a non-profit organisation that is dedicated to seeking technological solutions, sponsoring industry pilots, and fostering market and business developments.

The Eco system is to fulfil the following roles:

- A middleware that facilitates interoperation through supporting services such as authentication, billing, payment, and directories.

- An object-oriented development environment that encourages the re-use of EC modules.

- An industry roadmap and interoperability example that promotes open standards and helps technology vendors communicate with end users about product features.

The Eco System will consist of an extensible object-oriented framework as building blocks of EC applications. A Common Business Language lets application agents communicate using messages and objects that model communication in the real business world. A network services will insulate application agents from each other and from platform dependencies, while facilitating their interoperation.

Eco system's frameworks fall into four categories:

Internet Market services serve the Internet market. These are vertical markets of closely aligned businesses, such as real estate.

Business services include generic business processes and applications common to multiple Internet markets, including retail and business-to-business.

Marketware is a special subclass ot these services that links buyers and sellers. Its modules serve as building blocks to implement a variety of value added markets and market services, including matchmaking, negotiation buy-sell brokering, directories, and aggregation.

Commerce services are basic EC services, such as digital wallets for individuals and companies. Advanced commerce services will include secure multimedia mail, smart-card-based security and payment, digital-content delivery, application billing and accounting, transaction management, and agent management.

Network services enhance the performance, reliability, and security of the Internet to accommodate mission-critical business needs, incl. QoS management, IP multicast, delivery receipts, smart firewalls, authenticated packets.

Internet Market Service	Internet Market Service	Internet Market Service

Business Services

Commerce Services

Network Services

The Eco sytem is based on CORBA 2.0 which includes the IIOP which Netscape Communicator will support. Eco will also work with HTTP, HTML, and Java. Eco applications will be network-accessible services provided by agents. Eco agents respond to CBL messages form others agents and to HTTP requests from browsers. Interoperation will be achieved in many ways, incl. de-facto standards implemented in Java. Protocol negotiation, gateways, and mediators will provide semantic interoperation.

4 The SECCO Support Environment for Electronic Commerce

FOKUS is an active member of the OMG. Its developments, therefore, will be closely aligned with the EC-DTF activities. The CommerceNet approach will be of particular interest because of its interoperability and integration framework supported by most of the major EC platform providers.

4.1 The Architecture of SECCO

The architecture identifies the SECCO components and categorises them according to a structure which reflects necessity and complexity of usage. It includes the following major building blocks:

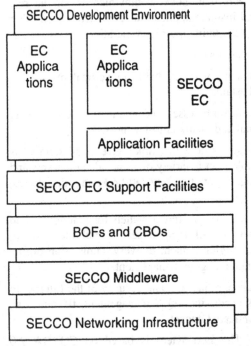

A **Networking Infrastructure** in an Intranet, Internet, or global environment. SECCO will build on existing technology, participate in trials on innovative technologies, and validate/ demonstrate developments in inter-connected networks.

The **SECCO Middleware** concerned with the support of communication, collaboration, and distributed object-oriented computing. Interoperation of EC applications, business objects, and SECCO facilities relies on that middleware. Some of its components may be extended or specialised so that they provide more enhanced functionality for specific use in EC applications. SECCO will use existing technologies, integrate available products, support interoperability, and fill in gaps by own developments and products.

The **Business Object Facilities and Common Business Objects** will be integrated as a platform providing support for business processes in general. SECCO will primarily use existing approaches, integrate available products, and probably fill in gaps by own developments.

The **SECCO EC Support Facilities** providing support services and facilities satisfying the specific requirements of Electronic Commerce in open, (global) electronic markets. SECCO will consider requirements identified by OMG, OSM and CommerceNet, and will develop its own components in line with the relevant activities, described above. Functions such as selection, negotiation, contract management, subscription, accounting, service and session management, as well as advertising, browsing, and brokerage are candidates for development. SECCO developments will integrate existing FOKUS (7) and IKV (8) technology and products and produce advanced services and facilities.

The **SECCO EC Application Facilities** are major building blocks of EC applications, EC applications of their own, or both. Examples are the multimedia communication and the media-on-demand facilities which may be used as building blocks for exchanging profiles, catalogues, or contracts in a trade scenario – or they may serve as EC applications of their own, i.e. an enterprise may provide the services to customers for profit making.

The **SECCO Development Environment** provides models, methods, and tools for the life-cycle support of components and applications implementing the rules and prescriptions of the SECCO architecture.

4.2 Available SECCO Components

SECCO developments can be based on a set of enabling technologies, know how, and products which are to fill-in the categories identified by the architecture.[3]

4.2.1 Network Infrastructure

Global Networking Testbed. FOKUS is closely co-operating with partners involved in collaborative projects, world-wide trials, and demonstrations for broadband communication and information services. These activities facilitate the access to a global broadband networking testbed. A further infrastructure component is the IMSS (Internet Mobility Support System) which provides a hard- and software platform based on Sun-SPARC systems and PCs with various implementations of mobile IP running under Solaris and WindowsNT.

4.2.2 Middleware Services and Facilities

Secure Middleware, is an open platform comprising generic security services based on secret-and public-key cryptography, including identification and authentication of human users and objects, authorisation and access control, security auditing, security of communication, non-repudiation services, and the administration of security information.

Stream Support for audio/ video data flows is provided by a Distributed Processing Environment which is a CORBA-based platform compliant to the TINA-C (9) specifications for modern telecommunications and open distributed processing.

Trader, an OMG/ ODP-compliant product which can be used in directory, information, and brokering services and applications for pre-sales support, establishing contacts between suppliers and customers, and advanced searches on suppliers, products, goods, services, etc.

---

[3] Some of these components are still under construction but first prototypes are available in any case.

Type Manager, a service which supports the structuring and management of information objects which can be used to describe products, documents, profiles, contracts, service offers etc. Type hierarchies and domains are administered. A browser GUI is available.

OMG-MAF products conformant to the OMG Mobile Agent Facility (MAF) specification are available by the end of 1997. The FOKUS Competence Centre on Intelligent Mobile Agents is currently working on the OMG-MAF specification in co-operation with Crystalize, General magic, IBM, and The Open Group.

Notification Service, a service that provides event filtering capabilities as an alternative to event polling mechanisms. It enables event consumers to express interest in events that are filtered by type and content, and enables event suppliers to provide filterable events to the service.

4.2.3 SECCO EC Support Facilities

Service Invocation, Profile and Session Management are supported by the Distributed Processing Environment Y.DPE (7) which is a CORBA-based platform compliant to the TINA-C service architecture. The architecture specifies – among others - a *service access session* and a *service session* which allow mobility and personalisation based on profile management.

Subscription and Accounting Service, two services compliant to the TINA-C service architecture which support the management and control of the access and the use of any communication or information service. Subscription provides facilities to use services individually. It controls the use of services and the access to service specific capabilities. Accounting handles application specific charging mechanisms and provides customer specific billing.

Federation Manager and Policy Manager, two components which support the management of agreements between organisational and technical domains and the management of rules and policies governing relationships and interactions between components. For the application in electronic commerce transactions they will be enhanced and elaborated based on standard negotiation and contract facilities.

4.2.4 SECCO EC Application Facilities

Multimedia Communication Service, a service that supports audio-video conferencing of an arbitrary number of participants. Enhancements for personal mobility are available.

Media-on-Demand Service, provides access to multimedia information based on both DAVIC recommendations and Internet standards. Application are Video-on-Demand and Electronic Program Guide in a broadcast environment. Applications may include native MHEG-5 clients and servers with HTML-MHEG gateways as well as WWW browsers using Java applets or plug-ins for the presentation of MHEG-5 objects.

WebStore, a managed WWW-based secure multimedia document store which provides a comfortable customer interface and advanced management facilities for private and public information. It can be used stand-alone or as a building block for advanced information services.

Open Trust Centre, a service which is being developed based on international standards such as X.509, LDAP, and PKCS 11 for certification, public certificate server, and SmartCard issue. Key recovery services are also provided.

4.2.5   Development Environment

Service Creation Environment, an environment developed for the creation of re-usable service components and the construction and configuration of services out of these components. The environment includes a set of tools based on models and languages, primarily supporting the computational specification by means of ODL, CORBA/ IDL, and SDL for behaviour. Object-oriented design is supported by use-cases and OMT.

The development environment is based on the Y.SCE (8) which comprises a framework for analysis, design, and implementation of telecommunication services based on the RM-ODP. The different phases of the software lifecycle are supported by user interfaces to be used by system analysts and architects as well as by application programmers. Tool-supported development of ODP viewpoint models and specifications is one of the main features of the Y.SCE.

5   An Application Example

5.1   The scenario

Imagine a software component market where data structures together with sets of software modules required for processing these data are distributed among users of different categories and locations. Such a market may for example include geographical data, such as road maps, weather-charts, etc.

In such a market we may at least identify the following players involved:

- The *providers* of data and software which produce new components, update/ modify/ replace/ delete existing ones, and deliver components according to negotiated contracts or spontaneous requests to individuals or groups of users.

- The *users* which receive the deliveries and use them continuously or for some periods of time.

- The *customers* which have negotiated with the providers the delivery of components to users and which have to pay for their users.

- A *market support system* which administers the market and provides the market infrastructure services described below.

5.2   SECCO support for business processes and transactions

Global, broadband communication. We assume that the players involved have access to (global) communication facilities such as provided by the SECCO *network* *infrastructure*. In this way they will have global broadband facilities available and partners may be distributed world-wide.

We further assume that they reside on the Y.DPE *distributed processing environment*, and thus, they will have stream support for multimedia communications in point-to-point and multi-point settings.

Initial contact and information exchange. Providers are able to advertise components via the OMG trader in a federated environment, i.e. several provider and customer domains may be federated through interworking traders and associated *federation managers* which handle advertising, information exchange and trading across customer and provider domains. Targeting advertisement to specific groups of customers, e.g. airlines for weather charts, and navigation through the provider space may be additionally supported by customer, user, and provider profiles which describe specific capabilities and interests. In our scenario, a profile of a weather chart provider may contain the update rate of satellite data, a customer profile may contain information about the mobility of its users, and a user profile may contain information about the data presentation facilities available. Such profiles are administered by the *profile manager*.

Pre-sales and post-sales support. Catalogues and details about available products, including multimedia data, may be exchanged between the component providers and their customers by means of the *multimedia communication and the media-on-demand* services. Guidance on product use can also be implemented by these multimedia services, e.g. by downloading of audio-visual courses or by contacting experts for advise.

Negotiation and contract making. An important aspect of negotiations is to harmonise conflicting policies and to reach agreements that may be formalised by contracts. In our scenario such agreements may include organisational aspects, such as cost reduction for long-term contracts, or technical aspects, such as the target environment for component downloading. Policy administration is performed by the SECCO *policy manager*. Authorisation of negotiators is supported by the *Open Trust Centre*, and critical information is exchanged via the *secure middleware*. The *Webstore* provides storage and management facilities for the administration of documents, such as contracts, profiles, etc.

Subscription and session control. Prior to sales and delivery, customers may subscribe to the providers services, e.g. the delivery of components and the periodical or occasional update of such components, according to the contracts and agreements negotiated. The SECCO *subscription service* is able to control the access and use to a delivery and updating service, including access to specific capabilities, such as timely delivery or immediate notification of data changes. Component downloading and usage constraints can be organised as service sessions which can be managed by the *Y.DPE service session manager*. Accounting of and charging for the use of components according to negotiated tariffs and customer specific billing can be handled by the SECCO *accounting service*.

Sales and delivery. The delivery and use of the software components, the downloading and processing of data, is supported in our scenario by various SECCO services. First of all, the storage and administration of the components at the provider's site can be supported by the *Webstore* which provides facilities for the structured and user oriented storage and management of information. For example, groups of component users can be defined, associated contracts and agreements administered, notifications of component changes can be generated, and periodical data updates can be performed. Information transfer, again, can be supported by *the multimedia communication* and *media-on-demand* services. Critical information transfer, e.g. military data, may be additional supported by the services of the *secure middleware*.

6 Conclusion and Future Plans

The example briefly described in the previous section has outlined that existing SECCO components already provide a useful spectrum of EC support services. The components have been developed according to the CORBA and TINA specifications, and thus are interoperable based on the standards provided by the related architectures and vendor independent. However, some of the components, such as the policy and federation manager for example, need to be validated for its use in electronic commerce by trial applications, particularly in widely distributed environments. Others must be reconsidered to make them more effective and more customisable for EC applications. Finally, we would like to offer dedicated packages, e.g. for brokerage and retailing, with components tailored to specific technical environments and fit for customisation to specific organisational and application environments. We also admit that SECCO has little support for electronic payment and related issues.

Further tasks are the integration of Business Object Facilities and Common Business Objects. Interoperability with major EC platforms coming out of the CommerceNet consortium is another important issue.

7 References

(1) J.M. Tenenbaum, T.S. Chowdry, and K. Hughes. Eco System: An Internet Commerce Architecture. IEEE Computer. Vol. 30, 5, May 1997.

(2) BOF/MOF/OA&D Boundary Guidelines, Draft 5, OMG Document Bom/97-04-01

(3) http://www.cordis.lu/esprit/src/ecomhome.htm

(4) http://www.osm.net/ec-dtf/model/components.html

(5) Cunningham et al., ACTS Project OSM, Reports, Feb. 1997

(6) http://www.commerce.net/eco/index.html

(7) http://www.fokus.gmd.de/

(8) http://www.ikv.de/pright.html

(9) http://www.tinac.com/

A Bridge for Heterogeneous Communication between CORBA and DCE

Dong Jin Kim, Han Namgoong and Young-Chul Lew

Distributed Processing Section
Electronics and Telecommunications Research Institute
Yusong P.O. Box 106, TAEJEON, 305-600 S. KOREA
Email: djkim@etri.re.kr, URL: http://hanuri0.etri.re.kr/djkim

Abstract. This paper considers a bridge which can be used for the heterogeneous communication between Common Object Request Broker Architecture(CORBA) and Distributed Computing Environment(DCE). A bridge can overcome the heterogeneity of different distributed middleware platform. The key point in creating bridges is the design and implementation of the mapping function and Interface Definition Languages(IDL) translation enabling communication between them. The main work addressed in this paper involves the design and implementation of a couple of technologies: First, IDL translation tool which translates an existing DCE interface definition into an equivalent CORBA interface definition. Second, a bridge code generation tool which automatically generates a CORBA-to-DCE bridge, based on a DCE interface definition.

Keywords: Distributed Computing Environment(DCE), Common Object Request Broker Architecture(CORBA), Interface Definition Languages(IDL), translation.

1 Introduction

Currently, Open Software Foundation(OSF)'s Distributed Computing Environment(DCE) and Object Management Group(OMG)'s Common Object Request Broker Architecture(CORBA) are two major distributed system infrastructures. Both of them provide distributed interprocess communication, however, the communication mechanisms are quite different and there is no compatibility between them. DCE was developed in a conventional procedural-oriented C programming environment. On the other hand, CORBA was developed in an object-oriented C++ programming environment. These two different development environment effect on the user's application programming. DCE evolved from the integration of the strongest technologies based on existing software. However CORBA was driven by a specification which was discussed among the major computer vendors. The actual commercial implementation of CORBA was left to the vendors, so there are some incompatibilities among CORBA platforms.

DCE provides security, a naming service, and remote procedure calls(RPC) which is a basis for communication among heterogeneous systems. DCE application programming requires significant knowledge of DCE technology. Basically, the DCE server interface has to make multiple calls to register itself with the name service, and a client also has to make multiple calls to find a required interface. In contrast, CORBA application programming is simple and transparent. The IDL used by CORBA and DCE are incompatible. In addition, compatibility does not exist neither in the generated stub nor in the skeleton code. The key point in creating bridges is the design and implementation of the mapping function and communication between them via translation of IDL. Results from the DCE server can be returned to the CORBA client following the operation. This paper is organized as follows: The next Section contain the objectives of the bridge. In Section 3, this operation of a bridge was presented. Section 4 provides the IDL translation mechanism in detail, and the implementation of bridge in Section 5. Finally, the paper concludes in Section 6.

2 Objectives

The interoperability between CORBA and DCE involves communication between two heterogeneous distributed systems. Both OSF/DCE and OMG CORBA utilize an RPC type interoperation mechanism, however, the syntax and semantics of communication are significantly different. The characteristics of DCE include robustness and well defined mechanisms for commercial applications. CORBA provides an object-oriented language interface. Bridges between a CORBA client and DCE server could be the most valuable distributed computing resource, not only for overcoming incompatibility between CORBA and DCE, but also for the increased availability of accessing DCE server resources which already has been built. Basically, vise versa, a DCE client and CORBA server architecture is technically possible, however, there are no such application requirement.

The CORBA-DCE bridge cannot provide the functionality beyond which is common to each system. For instance, although the bridge may be used as a communication tool between CORBA and DCE, CORBA clients cannot access or use any of the security features which are provided by the DCE server, and vice versa, the Dynamic Invocation Interface cannot be provided by the DCE server for CORBA clients. Only the common set of functionality can be used on the current bridge.

The developed bridge can extend the multimedia application area such as VoD(Video on Demand) application using CORBA, and Java ORB enabled Web client can access the data which is located in DCE server domain.

3 Bridges

The operation of a bridge is a mixture of server and client operation. According to these roles, a bridge has a server-side interface which accepts requests from a client process, and a client-side interface which makes requests for a server process. The server-side interface is defined using the client's IDL, while the client-side interface is defined by using the server's IDL. Figure 1 shows the operation of a generic bridge. The numbers represent a chronological order of events, each of which is described below:

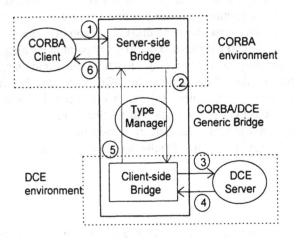

Fig. 1. The operation of a generic bridge

1. Bridge receives an operation invocation from a client in the CORBA domain.
2. Bridge transforms the parameters to equivalently typed parameters from the DCE domain.
3. Bridge invokes the operation in the DCE domain, which is equivalent to the original operation as invoked in (1) above.
4. The server in the DCE domain performs the operations requested by the invoking client and returns them to the bridge.
5. Bridge transforms any output or return parameters from the DCE domain, to equivalently typed parameters from the CORBA domain.
6. Bridge returns the output parameters and results to the client.

To make a generic bridge, both of the IDL files are required and compiled by their respective IDL compiler. The IDL compilers produce stub code which, among other functions, marshals and unmarshals the input and output arguments of each invoked operation. Intermediate bridging code resides between the two stubs, and translates values between CORBA and DCE. Three kind of bridges are identified, each of which are described in the following subsections.

In a static bridge, the set of operations which can be invoked by a client is fixed at compile time. Both the client side interface and server side interfaces of a static bridge are defined before compile time, using their respective IDLs. The bridging code is designed to only provide bridging for the operations defined by the client and server side interfaces. If the composition of the client and server application is changed, a new static bridge must be implemented and compiled at each time.

On-demand bridges can be considered an extension of static bridges. A bridge factory automatically generates the source code for a static bridge. In this paper, based on the DCE interface definition, bridging code and a CORBA side IDL is automatically generated and linked to the client side interface and server side interface stubs. Since the bridge code must translate values between CORBA and DCE, the mapping rules for the respective types are coded into the bridge factory. The invocation of the bridge factory can be part of the binding process which would make the creation of the bridge transparent to a client. The operation of on-demand bridge is shown in Figure 2.

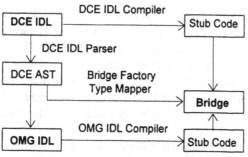

Fig. 2. On-demand bridge

Dynamic bridges support any IDL interface; the client/server interface types do not have to be known at compile time. To enable this dynamic nature of the bridge, the concept of a Dynamic Invocation Interface is required. The CORBA Dynamic Invocation Interface is specified in the CORBA architecture and specification[1] and is implemented in most of the commercially available ORBs. However, DCE does not provide such a facility.

When interface specifications are likely to change or services are frequently created, the static bridge approach becomes a programming bottleneck. This problem can be overcome by replacing static bridges with on-demand bridges. On-demand bridges relieve the programming bottleneck by automating the task of bridge generation. The bridge factory would be invoked and would generate a bridge, based on the DCE interface.

Dynamic bridges are more flexible than either static or on-demand bridges, since they do not need to be re-implemented for each new interface definition. However, dynamic bridges are more complex. Furthermore, the non-transformable concept difference between CORBA and DCE make it hard to implement. Non-transformable concept will be discussed in Section 4.

The other approach is using Environment Specific Inter-ORB Protocol(ESIOP), the DCE Common Inter-ORB Protocol(DCE-CIOP)[1]. DCE-CIOP messages represent OMG IDL types by using the Common Data Representation(CDR) transfer syntax. DCE-CIOP message headers and bodies are specified as OMG IDL types. That is, DCE application servers can be called from CORBA client through the specified CIOP without any addtional bridges. However, the actual implementation could be so complicated. Because, every transfer function and type mapping parts have to be implemented in CORBA side. Most of the DCE-CIOP specification was focused on the location of server objects and their invocation, which is not enough to implement and the actual part of forwarding requests, translation of data structure, type mapping and finally retrieving results from server side. So, in this paper, for the simplicity, we adapted the alternative way to overcome the heterogeneity between CORBA and DCE by using bridge.

4 Interface Translation

One of the most important things in making bridges is the transformation of interface types between CORBA and DCE[3]. So current bridge factory for the on-demand bridge only takes the interface definition from the DCE domain. The bridge factory's first task is to generate a corresponding interface definition for the CORBA domain.

Both bridge interfaces must be semantically equivalent, even though the client side interface is describe in terms of CORBA IDL and the server side interface is described in terms of DCE IDL. For this goal to be achieved, the bridge factory must include a set of mapping rules for the respective types.

The DCE type language supports a similar but not identical array of types than is supported by CORBA IDL. In particular, three classes of DCE-to-CORBA transformations have been identified.

1. Direct transformations: This class of transformation involves the syntactical transformation of equivalent concepts such as most basic data types, arrays, structures, unions and operations.
2. Indirect transformations: This class of transformation involves concepts which have no syntactic equivalent in the other IDL, but which can be adequately represented by a combination of other constructs. This includes DCE's pointer and CORBA's sequence, and exception types.
3. Non-transformable: In both IDLs there are concepts which are specific to the particular middleware environment, e.g., CORBA's context expressions or DCE's pipe. These are consequently not expressible in the other IDL. This problem have to be overcome by using customized mappings. That is, non transformable concepts should be mapped to extra-IDL concepts that are provided by particular services. These include user defined services and OMG's CORBA Services: Common Object Service Specification[2].

4.1 Direct transformations

There exist a subset of DCE and CORBA IDL types which have the same definition, and thus exactly equivalent. Direct transformation has been identified for the IDL scalar types shown in Table 1.

In addition to the scalar types, a number of DCE IDL structured types have direct CORBA equivalents.

Fixed length arrays A fixed array has constant index values for its dimensions; the length of a fixed array is set at compile-time. DCE fixed array are directly equivalent to CORBA arrays.

Table 1. Direct DCE IDL to CORBA IDL translations

DCE IDL type	CORBA IDL type
boolean	boolean
byte	octet
char	char
short	short
long	long
unsigned short	unsigned short
unsigned long	unsigned long
float	float
double	double
error_status_t	unsigned long

Enumerated types   DCE IDL enumerated types directly correspond to their equivalent CORBA IDL. For example, the DCE type definition

```
typedef enum {PUSH, POP, TOP} stack;
```

is equivalent to the CORBA IDL definition

```
enum stack {PUSH, POP, TOP};
```

4.2 Indirect transformations

There exist a number of CORBA IDL types which have no equivalent DCE IDL types, however they may be represented indirectly using DCE IDL.

Integer types   A number of DCE IDL integer types have no direct equivalent CORBA IDL. It is currently proposed that these DCE types be represented using either CORBA octets, or fixed arrays of octets. Table 2 shows the proposed translations.

Table 2. Indirect integer translations

DCE IDL type	CORBA IDL type
small	octet
unsigned small	octet
hyper	octet[4]
unsigned hyper	octet[4]

Pointers DCE pointers can be qualified by the attributes string, ref and ptr. Depending on the qualifying attributes they have different semantics.

DCE string pointers are identified by the string attribute. String pointers contain the address of a null-terminated character string and are equivalent to fixed or varying length CORBA IDL string types. For example, the DCE IDL type declarations

```
typedef [string] char foo[MAXSTRING]; /* Fixed length */
typedef [string, ptr] char *bar;        /* Varying length */
```

are converted to the following CORBA IDL type declarations

```
typedef char foo[MAXSTRING];   or
typedef string<MAXSTRING> foo;
typedef string bar;
```

DCE full pointers have all the capabilities usually associated with C pointers. CORBA does not include an implementation of pointers. Full pointers are converted to CORBA sequences of length one, containing a transformation into CORBA IDL of the data structure that was originally pointed to. A sequence of size 0 refers to NULL while a sequence of size 1 contains the equivalent of what the DCE full pointer was pointing to. For an instance, the DCE IDL type declarations

```
typedef [ptr] long *distance;
```

is converted to the CORBA IDL type declaration

```
typedef sequence <long, 1> distance;
```

The above type mapping must be extended when DCE full pointers to recursive or self-referential structures are considered. The simplest example of a recursive structure is a linked list. Each element of the linked list contains a pointer to another element of the same type. In general, DCE full pointers are equivalent to the CORBA unbounded sequence.

For example, assuming a definition of the datatype type exists, the following DCE declaration

```
struct list_t {
    datatype element;
    list_t   *next;
} list;
```

is converted to CORBA unbounded sequence

```
sequence <datatype> list;
```

The above approach for lists can be generalized for trees. For example, consider following DCE IDL structure defining a tree:

```
structure binary_tree_t {
    datatype element;
    [ptr] binary_tree_t *left;
    [ptr] binary_tree_t *right;
} binary_tree;
```

This is converted into CORBA structure

```
structure binary_tree_t {
    datatype element;
    sequence <binary_tree_t,1> left;
    sequence <binary_tree_t,1> right; };
```

DCE reference pointers provide the distributed application programmer with a pointer which as restricted semantics, but which incurs less overhead at run-time. The server cannot allocate new memory for the client's reference pointer during a remote procedure call, and the client cannot pass a reference pointer parameter as a NULL value.

OMG IDL lacks a concept of pointers. Thus reference pointers to the data structures are mapped to the data structures they point to. For example, the DCE IDL type declaration

```
typedef [ref] long *distance;
```

is converted to the CORBA IDL type declaration

```
typedef long distance;
```

Variable length arrays DCE IDL provides two types of variable length arrays. The advantage of variable length arrays is that only the portion of the array that is being used required to be marshaled and transmitted during and RPC communication.

DCE conformant arrays are a type of variable length array which use a dimension variable to specify the array size at run-time. As the array size can vary over time, DCE conformant arrays are equivalent to CORBA unbounded sequences. Following example shows the equivalent IDL declaration between DCE and CORBA. The DCE IDL declaration

```
[size_of(size)] long foo[*];
```

is equivalent to the CORBA IDL declaration

```
sequence<long> foo;
```

DCE varying arrays use variables to represent subset portions of the while array at run-time. However, the subset of the array does not need to start at the first element, as is the case with conformant arrays. It is proposed that DCE IDL varying arrays be converted into CORBA IDL *structs*. For example, the DCE IDL varying array

```
[first_is(var1),
 length_is(var2)] long nos[MAXSIZE];
```

is equivalent to the CORBA IDL struct

```
struct nos {
     string lower;
     string uppermax;
     boolean upper;
     long nos[MAXSIZE]; };
```

Structures and unions Fixed length structures are structures whose length can be determined at compile time. Fixed length structs do not therefore include structs whose fields contain strings, pointers or references, varying length arrays, sequences or typedefs to any of these types.

Fixed length DCE IDL structs can be converted directly to equivalent CORBA IDL structs, since each field can be converted to its CORBA equivalent. For example, the DCE IDL struct

```
struct st1 {
     long v1;
     byte v2; };
```

is equivalent to the CORBA IDL struct

```
struct st1 {
     long v1;
     octet v2; };
```

In general, variable length DCE IDL structs can be converted to CORBA IDL structs, based on the rules for variable length types such as strings and pointers. For example, the DCE IDL struct

```
struct tree {
   long v1;
   byte v2;
   [ptr] char *v3;
   [ptr] struct tree *left;
   [ptr] struct tree *right; };
```

is equivalent to the CORBA IDL struct

```
struct tree {
   long v1;
   octet v2;
   sequence<char> v3;
   sequence<struct tree> left;
   sequence<struct tree> right; };
```

Both DCE and CORBA IDL use discriminated unions, where the discriminator provides a switch that indicates what type is currently being stored in the union. The above rules for DCE IDL to CORBA IDL structure conversion apply to union conversion as well. For example, the DCE IDL type definition.

```
typedef union switch (long var_name)
    identifier {
        case 1: long  x;
        case 2: short y; } id;
```

is equivalent to the CORBA IDL definition

```
union identifier switch (long) {
    case 1: long  x;
    case 2: short y; };
typedef identifier id;
```

Interface headers DCE and CORBA use a different syntax for their respective IDL interface definitions. DCE interface headers may contain imports, a unique interface identifier, a version number and various attributes.

DCE IDL **import** statements are syntactically equivalent to CORBA IDL **#include** statements. Besides the syntactic conversion of the *imports* into *includes*, there is the question of whether the imported interfaces should be translated by using the OMG IDL inheritance in order to avoid scoped naming problems with identifiers in the imported interfaces.

Operation results are converted to their equivalent types in OMG IDL. If the DCE operation execution attribute **maybe** is specified and the return type is **void**, with no output parameters then the equivalent CORBA declaration of the operation uses the **oneway** attribute.

The DCE **in, out** and **in,out** operation parameter attributes are equivalent to CORBA **in, out,** and **inout** parameter attributes. For example, the following DCE IDL operation definition

```
long Foo([in] char p1,
         [out] long *p2,
         [in,out] short **p3);
```

is equivalent to the CORBA IDL operation definition

```
long Foo(in char p1,
         out long p2,
         inout sequence <short> p3);
```

4.3 Non-transformable concepts

There are a number of DCE IDL types and concepts for which there appears to be no CORBA IDL equivalent.

1. DCE **pipes** are not represented by an equivalent or similar concept in OMG IDL. However, OMG's Telecom Domain Task Force is currently preparing a RFP(Request For Proposals) for streams[7]. OMG's streams are expected to cover the functionality of DCE pipes.

2. The DCE operation execution attributes **idempotent** and **broadcast** does not have equivalent CORBA IDL.

3. DCE IDL allows interfaces to be distinguished using a universal unique identifier(UUID). In addition, DCE can make use of version attributes when more than on version of an interface exists. The unique interface identifier and the version number can only be translated into implementation dependent pragmas[6] in OMG IDL, which may be discarded by the IDL compiler.

4. DCE provides a context handle attribute, which allows a client to maintain state on behalf of a server, between operation calls. There is no equivalent CORBA to DCE IDL context handles.

5. DCE allows programmers to define encoding and decoding operations on an interface. These operations can be used to flatten any idl type or structure to a **idl_byte** stream at run-time. The original type or structure can be consistently recovered from the byte stream. CORBA IDL has no equivalent operation specifiers.

6. A number of DCE interface attributes are implementation specific, including the **represent_as**, **enable_allocate** and **heap** attributes. CORBA IDL is purely a specification language, thus these DCE implementation specific features have no CORBA IDL equivalent.

Apart from the non-transformable concepts between CORBA and DCE, some unique environment features cannot be supported through the bridge. For instance,

1. Secure RPC service which is in the DCE server cannot be accessed from a CORBA client.

2. The DCE pipe functionality cannot be used from a CORBA client.

3. The Dynamic Invocation Interface cannot be supported on a DCE server for a CORBA client.

4. The Distributed time service, and Cell Directory Service are in same situation.

5   Implementation of the Bridge

The bridge factory takes only the interface definition from the DCE server domain. The role of the bridge factory is to produce a CORBA-to-DCE bridge, based on the DCE IDL client side interface definition. This role can be split into two tasks:

1. Generate a server side interface definition for the client's CORBA domain, based on the client side interface definition. The bridge factory will generate a server side interface definition described using CORBA IDL, and will base its generation on a client side interface definition written in DCE IDL. Both bridge interfaces must be semantically equivalent.

2. Generate the bridging code. This will take the request parameter values from each server side operation, and transform them to correct types for the equivalent client side interface operations. Any return values must be handled in reverse.

The bridging code can be divided into three parts.

Declarations The bridge factory which was made in this research assumes the existence of the client side interface definition. It is necessary to declare the local variables which will be used as actual parameters in the client side interface operation calls. Since these local variables are to be parameters to the client side interface procedure calls, they must be declared using types from the server's IDL. Parameters from the client request will be declared in each server side operation definition, using types from the client's IDL.

Assignments and transformations The generated bridge behaves as a proxy at run-time, receiving an operation invocation through its server side interface from a client in the CORBA domain, on behalf of a server in the DCE domain. To enable forwarding of operation invocations to the desired server, the values of any incoming CORBA request parameters must be transformed to the outgoing parameters of the corresponding DCE operation. If the types of the incoming and outgoing parameters are different, the parameter values must be transformed between the types.

House keeping The above two sections of bridging code are interface-dependent, and so must be generated for each new client side interface definition. In addition, all CORBA-to-DCE bridges require some standard code to perform certain house keeping functions. The interface-independent house keeping code performs such task as making the bridge ready for a CORBA client and binding to the DCE server.

On-demand bridges are created in response to request to bridge factories. The bridge factory operates in two stages.

1. In the first stage, the bridge parses the client side interface definition, and converts it to an abstract syntax tree(AST) representation.
2. Once the input interface definition is in this AST format, it is a more simple matter to traverse the AST, and generate the new interface and bridging code described above.

A bridge factory has been created using three tools, which is lex, yacc, and the term processor, Kimwitu[8]. These tools generates three programs which together implement the bridge's operational stages described above. These three programs generated are:

1. The parser, dce-idl2cr, which can parse a language(DCE IDL) file and convert the file to an AST representation, based on the language's grammar. The Kimwitu compiler generates the dce-idl2cr parser source code based on the description of the DCE IDL grammar. The source files produced by YACC and Kimwitu are compiled to produce the dce-idl2cr executable file as shown in Figure 3.

2. The code generator bridge factory, which generates two sets of C++ source code, based on the description of the AST representation. These are a generic DCE server side definition file, and produce an equivalent CORBA client side interface definition and associated bridging code.

The Kimwitu generated tools, dce-idl2cr and bridgeGen are to produce an on-demand bridge. The bridge factory will take a generic DCE server side definition file, and produce an equivalent CORBA client side interface definition and associated bridging code. The products of the bridge factory are compiled to produce the desired on-demand bridge.

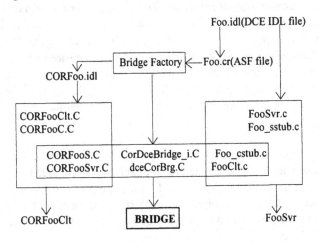

Fig. 3. Diagram of CORBA DCE bridge generation through the bridge factory

6 Conclusions

In this paper, the design and implementation of an on-demand bridge between CORBA client and DCE server has been addressed. The research on the interoperation between the two different distributed infrastructures has resulted in the definition of mapping from DCE IDL to OMG IDL, and vice versa, and in the implementation of corresponding tools. Applying these tools has been essential and effective to implement bridge code. The existing paradigm of interoperability is based on IDL descriptions and their translations, which are shared by both the CORBA and DCE environments.

The interoperation between CORBA and DCE is indispensable for an object-oriented client application environment which is simple, without the complication of a DCE environment. The result of this research can be regarded as a first step towards overcoming the incompatibility between two major distributed systems. Not only for the feasibility test of interoperability, but also for actual applications.

References

1. Object Management Group, *"The Common Object Request Broker: Architecture and Specification,"* Object Management Group(OMG), Framingham, MA., July 1995

2. Object Management Group, *"Common Object Services Specification, Volume 1,"* OMG Document Number 94-1-1, Object Management Group, Framingham, MA., 1994.

3. Vogel, A. and Grey, B., "Translating DCE IDL in OMG IDL and vice versa," *Technical Report #22*, CRC for Distributed Systems Technology, 1995

4. Brookes, W., Indulska, J. and Vogel, A., "A Type Management System Supporting Interoperability of Distributed Applications," *Proc. DSTC Symposium*, July, 1995.

5. Beitz, A. and Bearman, M., "An ODP Trading Service for DCE," *Proc. First International Workshop on Services in Distributed and Networked Environments(SDNE)*, pages 34-41, 1995.

6. Digital Equipment Corporation, *"DCE Application Development Reference,"* Digital Equipment Corporation, Maynard, Ma, 1992.

7. Raymond, K., "Streams for CORBA," White Paper, OMG, 1996

8. van Eijk, P. and Belinkfante, A., "The Term Processor Kimwitu. Manuel and Cookbook," University of Twente, The Netherlands, March, 1993

Distributed Connection Management for Real-Time Multimedia Services

Huw Oliver[1], Mark Banfield[2] and David Hutchison[2]

Abstract

We report on work in progress in the ReTINA project. A new architecture for supporting real-time multimedia services is being developed, implemented and tested. This new architecture is characterised by the application of distributed object principles to the control plane of the network. We show how this separation enables the maintenance of QoS estimates for the network and their use in routing and high-level admission control. At the application level we show how Winsock 2 can complement the architecture. Partial performance results for a restricted network are presented.

1. Introduction

The ReTINA project (ACTS AC048) has the goal of supporting real-time multimedia services over emerging broadband networks. It has used a distributed object infrastructure as the control plane for that network and the programme of work has covered aspects such as end system control, object models for streams, advanced computing services and software engineering tools for service construction.

The focus of this paper is the provision of Quality of Service (QoS) within such an architecture for applications with highly diverse and demanding expectations on traffic latency, bandwidth and reliability. We derive QoS estimates for use in routing and higher-order admission control. We believe that the architecture exhibits the dual advantages of logically centralized calculation together with truly distributed processing gains.

We describe our system architecture, engineering issues and implementation results so far.

2. Multimedia Services

The requirements of multimedia applications in terms of their traffic characteristics have been the subject of considerable study and are reasonably well understood. These requirements are usually expressed in terms of traffic parameters such as bandwidth, delay, jitter and error rate. The combination of these is referred to as the Quality of Service (QoS) required by the multimedia application. ReTINA follows the popular approach of adopting a contractual view of QoS. An application states the characteristics of its traffic together with a set of requirements it would like the

---

[1] Hewlett-Packard Laboratories, Filton Road, Stoke Gifford, Bristol BS12 6QZ, UK.
heo@hplb.hpl.hp.com
[2] Computing Department, Lancaster University, Bailrigg, Lancaster, LA1 4YR, UK.
[banfield,dh]@comp.lancs.ac.uk

60

network to provide. If the network is able to comply then these two descriptions form the basis of a contract between the application and the network services.

2.1 Quality of Service Specification

In order to establish a contract, a means of specifying QoS is required that is meaningful to the application in terms of its high level requirements and meaningful to the network in terms of physical switching technology. Example high-level requirements might be, "an MPEG1 encoded video stream" or "a PCM encoded POTS stream" or even, "good quality audio".

The switching fabric, however, will provide QoS based on traffic parameters such as bandwidth and error rate. A Connection Management Architecture (CMA) must bridge this gap. Our approach has been to introduce an intermediate "Generic QoS" service class. We accept requests from applications in high level multimedia terms, map these into generic QoS parameters and, at the lowest level, map the generic parameters into technology specific switch parameters.

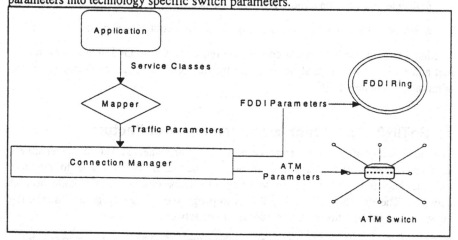

Figure 1: QoS Mapping

A further pragmatic requirement we imposed upon ourselves was motivated by the belief that the majority of multimedia traffic in the near future will be generated by Winsock 2.0 [3] applications. We describe how we achieved compatibility between our architecture and that of Winsock in section 7.

3. Network Services

The support we can provide for real-time services is ultimately constrained by the capabilities provided by the underlying network technology. We have used ATM for testing purposes because of its ability to support multiple traffic classes simultaneously and because it increases the negotiable QoS by making switch resources more configurable.

ATM uses a statistical multiplexing approach to the simultaneous support of multiple connections. Also, connections do not reserve the network resources they need for the entire duration of their lifetime. These efficiency-motivated aspects have two important effects in an ATM network:

- congestion is possible when excessive amounts of traffic contend for limited buffer resources (this results in lost cells)
- variable cell delays occur as the cells traverse the network. These are caused mainly by random queuing delays at each switch and multiplexer (the queues themselves are there to avoid excessive cell loss).

These phenomena lead to varying qualities of service on different routes through the network. Our approach to the problem has been to maintain estimates of the QoS capacities throughout the network in order to improve the network's ability to deliver a connection to an application with the requested QoS. It does this by using the QoS estimates

1. to decide on routes on a more informed basis;

2. as the basis for a high level connection admission control algorithm.

The latter is used partly to reject connection requests if their QoS demands cannot be met and partly to ensure that the existing traffic's QoS will not be disrupted by the admission of the new traffic.

4. ReTINA Connection Management Architecture

The large scale of today's networks and the interconnected nature of separately administered networks means that it quickly becomes impracticable to maintain accurate knowledge about all node and link states throughout the collection of networks. The ReTINA solution to this is to progressively aggregate, or abstract, the network state information into a hierarchy of subnetworks.

ReTINA's Connection Management Architecture (CMA) is characterized by a hierarchy of Connection Performers (CPs) acting over these subnetworks, until, at the lowest level, we reach open interfaces to switch resources. This hierarchy imposes a parent-child relationship between subnetworks. A child subnetwork is represented in the parent as a single node, with only its access points shown. The further up the hierarchy we go the more the information has been abstracted. One consequence of this is that there is necessarily a loss of information as we move up the hierarchy.

The ATM Forum PNNI standard [1] follows the same approach. Here the hierarchical aggregation is very similar but routing is source based and set-up initiation is signalled along the chosen route. The state changes in the links have to be periodically collected centrally. Link Status Update algorithms are introduced for addressing this. The CMA has one distinct advantage over this scheme when it comes to maintaining this hierarchically aggregated network state information. The requests for connection and disconnection will pass through the appropriate controlling object associated with the subnetwork to which the request was made. The decision whether to accept or deny the request is also made centrally within that subnetwork. Thus, the controlling object has direct access to the changes taking place in its subnetwork.

Another advantage of separating the control and payload channels can be seen in times of network congestion. In signalling based schemes such as RSVP [2], if the control messages are in-band they are likely to be lost at times of severe congestion - just when they are most needed. Suggested solutions include assigning separate virtual channels or the use of priorities for control messages.

In summary, we believe the ReTINA CMA does offer advantages for maintaining QoS status information but, as with other approaches, we are still necessarily operating with inaccurate information.

5. Quality of Service Management

We now present in more depth our approach to the use of that QoS status information. The low-level switch issues will be addressed in Section 6.

5.1 Hierarchical QoS-Based Routing

A ReTINA network is modelled as a number of Layered Networks. A Layer Network is made up of (partitioned into) subnetworks interconnected via links. Each Subnetwork, in turn, can be partitioned into smaller Subnetworks. At the lowest partitioning level, subnetworks represent individual network elements (e.g. ATM switches). This forms the network hierarchy (see Figure 2)

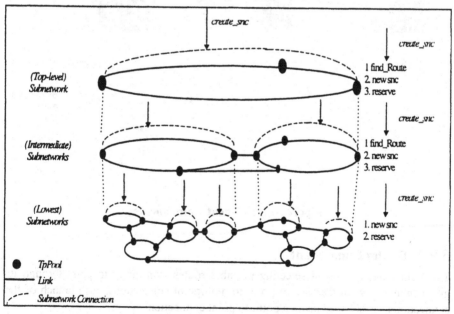

Figure 2: Hierarchical Routing

Routing under the Connection Manager is performed by a technique known as hierarchical routing. The process of routing involves making decisions at each subnetwork over which subordinate subnetworks to route upon based on whether those subnetworks can contribute to the goal of achieving the route. This is a divide and conquer solution, repeated until we get down to the subnetworks representing hardware switches.

Quality of Service is introduced into the routing model by choosing to use a subnetwork based not only on whether it contributes to our route but also whether it can provide the desired QoS characteristics.

Our approach is optimistic. We assume that the subordinate layers will be able to establish the desired route. We use an open transaction model where the actions to reserve resources are visible outside the transaction. If not, and it is quite possible that they will not because our QoS estimates are, after all, estimates, then we provide compensating actions to undo the effects of the transaction.

5.2 Functional Components

At each subnetwork level routing decisions are made by a QoS Agent. This consists of three components: a Router, a QoS Evaluator, and a Broker.

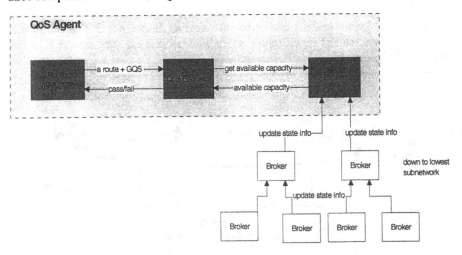

Figure 3: QoS Agent Components

5.2.1 Router Component

The Router has the job of selecting possible routes from point to point through its given subnetwork. In the case of point to multipoint connections, each branch of the connection is established using a point to point connection.

5.2.2 QoS Evaluator

The QoS Evaluator does two things:

1. It receives a proposed route through the subnetwork, and a QoS requirement (in terms of a Generic QoS Specification) from the Router component. It then decides whether the given route meets or fails the QoS requirement. This decision is made by querying the broker from the subnetwork about the QoS capacities involved in the route and by evaluating the effect of admitting the route.

2. It derives collapsed QoS estimates for a complete subnetwork as described in section 5.4.

Note that when a proposed route for a connection is being evaluated there are two things to consider. First, will the route be QoS-acceptable for the new connection. Secondly, will the newly accepted traffic break any of the existing QoS traffic contracts. To calculate the latter from scratch would be infeasible so we use a parameter called the "available bandwidth": the idea is that the QoS contracts will be maintained as long as the equivalent bandwidth of any new traffic does not exceed the available bandwidth.

5.2.3 Broker Component

The Broker is responsible for maintaining QoS capacity information of an associated resource (either a Subnetwork or Link). Initially we have chosen to store the available bandwidth, delay and error rate but this could be extended to include further QoS parameters, for example jitter. The broker hierarchy follows the network hierarchy and a broker is also responsible for initiating if necessary the update of its parent broker.

The Router, QoS Evaluator and Broker run in the same process for efficiency but the Broker runs in its own execution thread. This allows QoS estimate updates to occur as a low priority in the background but newly arrived connection requests have immediate access to the latest status figures.

5.3 Subnetwork Status Updates

The Broker at each subnetwork holds the QoS attributes for the nodes and links for that subnetwork. The QoS attribute values can be updated either by a sub-broker deciding that its status has changed sufficiently to "push" its new estimates up a level, or, by this broker deciding to "pull" or request estimates from the sub-broker. This happens, for example, at initialization.

It is necessary to get new estimates from the child broker because the topology of that subnetwork is unknown to this broker. One immediate question is how frequently these Subnetwork Status Updates (SSUs) occur. We have some constraints:

1. SSUs must not propagate all the way up the hierarchy to the root. That would cause the root to become a bottleneck.

2. SSUs must happen sufficiently often to keep the subnetwork capacity estimate accurate enough to get "good" routing decisions.

3. SSUs must not happen so frequently that they overburden the processing load.

Finding a solution to requirement 1 was not difficult. There are two reasons why we do not propagate updates to the root. First, the network capacity may be unaffected by the admission of a new connection (the new traffic may have been routed on a non-critical path) or second, the change in capacity may be too small or too transient to warrant an update to the parent broker.

The QoS estimates from the inferior brokers are cached in the parent broker since these QoS values are accessed frequently during route evaluation or QoS estimation. We use object proxies to make the caching transparent.

Requirements 2 and 3 led us to develop a scheme based on relative changes coupled with fixed time intervals. Optimising the parameters of this scheme is a piece of future work.

5.4 QoS Estimate Derivation

As described above, the need for a subnetwork to re-derive the estimate of its QoS capacity can be triggered either by a request from the parent or by the QoS Agent itself. In either case the activity is the same. The entire QoS-attributed topology has to be collapsed into the set of boundary nodes with QoS capacity estimates between those nodes.

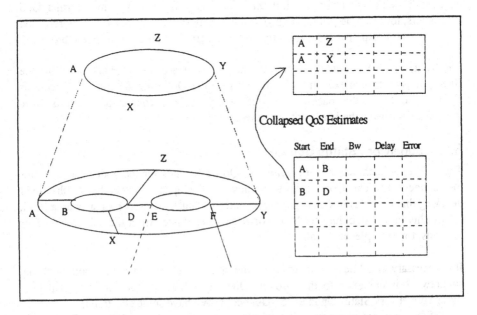

Figure 4: Deriving Subnetwork QoS Estimates

Initially we searched for heuristics or estimation algorithms that, while not exact, would quickly provide a rough estimate of QoS capacities across a subnetwork. We failed to find anything suitable and resorted to a complete analysis of the subnetwork using the different QoS parameters in turn as cost metrics for Dijkstra's shortest path algorithm. We were pleasantly surprised by the efficiency of this (not least compared to any remote object binding costs!). We are comfortable with these calculation costs, particularly remembering that they occur in a background execution thread.

Once we have a path (either from the Dijkstra algorithm or from the Router component) we can form an end-to-end value for the available bandwidth, delay and error rates.

6. At the Physical Level

Once the connection request reaches the leaves of the hierarchy, it is processed by EML-CPs. These are objects that can talk to the physical device (i.e. switch) directly. These separate switch set-up processes can proceed independently and in parallel.

Ideally we would like switches to export a control interface together with an advertisement of their current resources. In practice, today's switches do not do this and industry fora such as Opensig [5] exist to encourage the development of such open interfaces in future. The most promising candidate currently is GSMP [10].

The first issue, the lack of a control interface, can be tackled in two ways. Either we can use the management interface (specifically, SNMP) to set up connections, or, in some cases, we can manipulate the switch connectivity matrix directly. We discuss each of these approaches in the rest of the paper.

The absence of advertised switch resources was tackled by building simple mathematical models of the delay and error rates in the switch.

6.1 Switch Connectivity Matrix Manipulation

To overcome the above problem of no switch control interface, we built a Java Virtual Machine on the switch which hosted a program that directly manipulated the connectivity matrix. We were impressed by the use of the Java Virtual Machine on the different controllers to achieve platform portability and also the use of the JVM on the switch to build the server software. The use of 'Java everywhere' has allowed us to drop useful pieces of software onto the VM such as a Corba ORB. Without Java it would, for example, have been a more substantial piece of platform dependent work to embed an ORB on the switch.

We have experimented with different platforms for the controller (HP-UX and NT), using the same IIOP interface to the switch. Figure 3 shows two examples of the software pieces. The central block represents the switch while the right and left blocks show two alternative controllers.

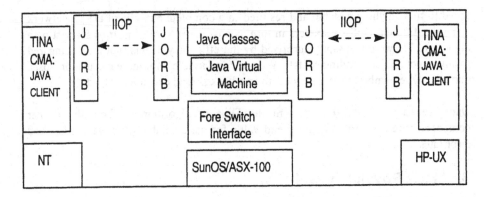

Figure 5: Java allows easy porting of the client software

This section describes a series of experiments and performance measurements carried out to assess the achievable performance of connection creation and deletion on an ATM switch. One of the comparisons we make in this section is between the use of SNMP as a switch control interface and a more direct control interface developed at HP Labs. The overall conclusion of these measurements is that SNMP performance can indeed be bettered by an order of magnitude by means of the alternative switching interface.

However, the performance measurements described in this section also provide input to other discussions, as listed below.

1. Performance of SNMP versus alternative link by link setup mechanisms.
2. Performance of C versus Java in I/O bound applications.
3. Performance of CORBA versus raw sockets.
4. Deployment of ReTINA CM computational objects.

Communication was done by means of two distinct mechanisms:
1. A CORBA/IIOP interface.
2. A raw interface based on TCP/IP sockets.

The connection server can be configured at start-up to provide either or both of the two interfaces, allowing clients to choose the means of communication. During measurements, the server was configured to only provide one of the two possible network interfaces at a time, although provision of multiple interfaces is unlikely to affect server performance.

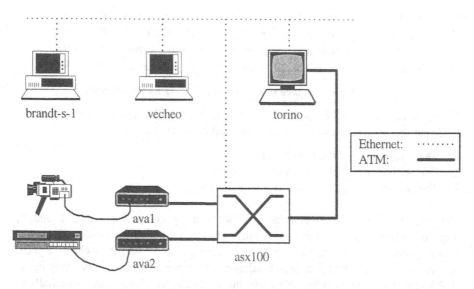

Figure 6: Source, Sinks and Connections

Measurements on pure CORBA solutions suggested that CORBA is too heavy-weight for communication with the physical switch. This realisation prompted implementation of a raw socket interface for comparison. However, to be plugged directly into the ReTINA framework, the server is required to provide a CORBA interface. Therefore, a hybrid solution was tested, using a CORBA proxy running on a fast workstation, and forwarding requests to the switch by means of the raw socket interface.

The ORB used was JORB (Java ORB) from HP, in alpha 3 release written entirely in Java, and providing a full CORBA 2.0 implementation.

	Network Interface	Client	Server	RPC/sec
1	JORB	torino/JDK 1.01	asx100/JDK 1.01	14
2	Socket	torino/JDK 1.01	asx100/JDK 1.01	106
3	JORB	torino/JDK 1.01	brandt-s-1/JDK 1.02	107
4	Socket	torino/JDK 1.01	brandt-s-1/JDK 1.02	597
5	JORB	brandt-s-1/Symantec	asx100/JDK 1.01	16
6	Socket	brandt-s-1/Symantec	asx100/JDK 1.01	121
7	JORB	brandt-s-1/Symantec	vecheo/Symantec	401
8	Socket	brandt-s-1/Symantec	vecheo/Symantec	1330
9	JORB	brandt-s-1/Symantec	brandt-s-1/Symantec	418
10	Socket	brandt-s-1/Symantec	brandt-s-1/Symantec	1818

11	JORB	brandt-s-1/JDK 1.02	brandt-s-1/JDK 1.02	159
12	Socket	brandt-s-1/JDK 1.02	brandt-s-1/JDK 1.02	998
13	JORB	brandt-s-1/Symantec	asx100/Kaffe	10
14	Socket	brandt-s-1/Symantec	asx100/Kaffe	105

Figure 7: performance measurements of Java based solutions.

As shown in row 1, JORB communication with the asx100 switch allows around 14 RPCs per second. This number is clearly limited by three main factors:

1. The performance of the switch itself.
2. The performance of the Java virtual machine.
3. The performance of the JORB implementation.

The impact of switch performance can be seen by comparing rows 1 and 3, respectively rows 2 and 4[3]. The performance gain from running the server on brandt-s-1, as opposed to asx100, is 7.6 using JORB, and 5.6 using sockets. Hence, performance of the switch is a clear bottleneck. That performance gain is larger in the JORB case must be contributed to the larger overhead of parameter marshalling, resulting in more time spent in the Java virtual machine.

The impact of virtual machine performance can be seen by comparing rows 9 and 11. Both rows reflect JORB performance with the client and server running on the same machine, in this case a 200MHz Pentium Pro PC. In row 9, the Symantec virtual machine is used, which includes a just-in-time (JIT) compiler. In row 11, the JDK 1.02 virtual machine is used. The performance gain from the JIT compiler is a factor 2.6 with JORB communication. Similarly, comparing rows 10 and 12 we get a factor 1.8 with raw socket communication. The difference in speed-up must again be contributed to the different times spent in the Java virtual machine, with the socket solution being the most I/O bound of the two solutions. In any case, similar speed-ups can in principle be achieved on the switch by applying a better Java virtual machine, although one was not available at the time the tests were made.

The impact of using CORBA is apparent from comparing any odd-numbered row to its following even-numbered row. For example, the difference between rows 1 and 2 is a speed-up of 7.6 from replacing JORB communication by socket communication.

6.1.1 A Hybrid Solution

Given the above measurements, it might be worthwhile to try to combine the best of the two worlds: the speed of plain sockets with the plug-ability of CORBA/IIOP. To that end, we tried to insert a proxy object between the client and the server: By means of JORB, the client makes calls to a proxy running on a fast PC on top of a fast Java implementation. The proxy then forwards the request to the real server, using the raw socket interface. Measurements are shown below:

---

[3] This assumes that JDK 1.01 and JDK 1.02 exhibit similar performance. This has not been tested.

	Client	Proxy	Server	RPC/sec
15	brandt-s-1/Symantec	brandt-s-1/Symantec	asx100/JDK 1.01	97

As can be seen, this approach does in fact combine the best of the two solutions. Combined with a fast JIT based Java implementation on the switch, around 200 connection creations per seconds should be comfortably within reach.

6.1.2 A Server Implementation in C

Finally, we implemented the server in C. The performance of the server written in C was measured in two different configurations: with the client connecting directly to the server using sockets, and with the client connecting via JORB to a proxy server, which then connects to the server using sockets. The C server does not provide a CORBA interface. Performance is shown below:

	Client	Proxy	Server	RPC/sec
16	brandt-s-1/Symantec	brandt-s-1/Symantec	asx100/C	251
17	brandt-s-1/Symantec	Direct (no proxy)	asx100/C	600

Comparing rows 15 and 16, we see that rewriting the connection server in C provides a speed-up factor of 2.6 in a proxy based solution. With the client calling the server directly by means of raw sockets, the C server provides a speed-up factor of 5, as can be seen by comparing rows 6 and 17. It should be noted that this speed-up would shrink to between 2 and 3 if a faster virtual machine could be used on the switch. However, this relatively small speed-up reflects the I/O bound nature of the server. Hence, if the computational complexity of the server was larger, a larger speed-up could be expected.

6.2 SNMP

Using a Smalltalk client and communicating with the switch by means of SNMP, around 50 connection creations or deletions per second was achieved. Assuming that Smalltalk exhibits performance similar to Java (this is ignoring the fact that many Smalltalk systems include a compiler), this number is comparable to row 17, providing around 600 connection creations per second. Hence, using a connection server instead of SNMP, an order of magnitude better performance can be achieved.

One improvement would be to use optimised implementations of SNMP on switches where the interface is being used for switch control. Another would be to use SNMPv2 which supports multiple get/set operations.

To be fair to SNMP, we should not forget that it is significantly more general than our protocol, applies to most or all switches for private networks (switches for public networks may support the CMIP protocol), and was furthermore not conceived as an alternative to connection setup by means of signaling. Nevertheless, although it can be improved upon, SNMP performance on the ASX-100 is probably sufficient for many purposes.

7. Winsock in the Connection Management Architecture

Winsock 2 is based on the socket paradigm introduced in BSD Unix (version 4.2 BSD in 1983), which was adopted for Windows in the form of Winsock 1 and 1.1. These socket interfaces, however, were developed for the slow packet switched networks of the early 1980s and lack the functionality required by the next generation of multimedia applications. Winsock 2 extends the socket paradigm to support new features such as QoS, multicast, multiple protocols and directory services.

7.1 Winsock Architecture

Winsock 2 has a modular architecture with well defined interfaces between components, enabling new services to be added easily. This is useful as new protocols can be installed and utilised by applications without having to redesign the application to use those new protocols.

There are three important components: name service providers, transport service providers, and layer service providers. Name service providers provide translation between naming schemes, for example Internet domain names to IP addresses. Transport service providers provide a means of transferring data over the underlying network technology. A layer service providers is an optional component that can be inserted between Winsock and other service providers, in the form of a protocol chain. It can be used to provide new service types that are based upon existing service providers. The Winsock 2 DLL acts as a switch, receiving operations from the application, and directing those operations to the relevant service provider.

7.2 Building a Broadband Multimedia Network with Winsock support

The ideal approach to supporting Winsock applications would have been to implement a new service provider stack that took advantage of our CMA. This would have been too large a piece of work for the project and so we used a commercial AAL5 stack.

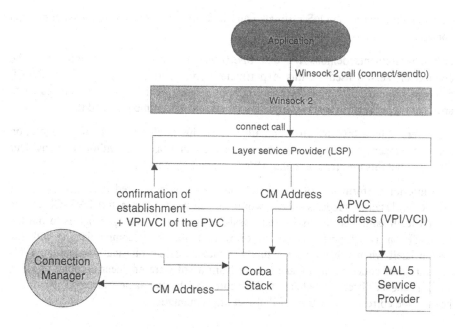

Figure 8: Using ReTINA's Connection Manager

In order to replace its use of UNI signalling we established connections with our Connection Manager and fooled the AAL5 stack into treating the connection as a Permanent Virtual Circuit (PVC). See Figure 8.

8. Conclusions

While Winsock is likely to be the most prevalent source of multimedia traffic, it requires some engineering to make it fit with new connection control systems. This is because it does not separate the notions of control and data transfer - implementations do both in the same monolithic stack.

Ideally we would re-implement the stack but, even with the temporary solution of getting the ATM stack to treat connections as PVCs (section 7.2), the system works well.

Our architectural separation of control and data gives us easy access to all control messages passing through the system. This makes the maintenance of QoS estimates feasible. We still have a lot of engineering to do to explore the parameter space and the scalability issues.

Fundamentally, the system works. We were able to switch multiple video and audio sources between multiple sinks (see Figure 6). We were able to improve route determination based on QoS estimates and it forms an effective high-level admission control.

Our measurements demonstrate that SNMP performance is indeed inferior to the connection server used in this experiment. On the test equipment used, SNMP achieves around 50 connection creations per second, which should be compared to around 600 per second achieved by the connection server implemented in C.

However, only a few switches actually allow the implementation of a connection server by means of an open switching interface such as used within HP Labs. Our ability to put this interface onto the ASX-100 is an exception.

The inferior performance of the switch hardware as compared to a modern PC or UNIX workstation suggests that it would be unwise to run a full EML-CP on the switch itself, although this is indeed possible. A more viable approach is to run the EML-CP on a high-performance PC or workstation, only communicating with the switch itself when necessary to actually create or delete physical connections. This separation of concerns also makes sense from a software engineering point of view, since it allows different switches to be controlled by the same (powerful) controller hosting the corresponding EML-CP object implementations.

9. Acknowledgements

The Java and C performance figures were measure by Soren Brandt of HP Labs. The performance of connection creation and deletion by means of SNMP was measured by Matthieu Goutet of HP Labs. The ReTINA CMA and QoS Agent design was a joint effort of the ReTINA partners.

10. References

[1] R. Cherukuri, D.Dykeman (eds.), and M.Goguen (chair). PNNI Draft Specification. ATM Forum 94-0471, November 1995.

[2] R.Braden (ed.), L.Zhang, S.Berson, S.Herzog and S.Jamin. Resource reSerVation Protocol (RSVP) version 1, functional specification (draft-ietf-rsvp-spec-13.ps). INTERNET-DRAFT, Internet Engineering Task Force - RSVP WG, July 1996.

[3] "Windows Sockets 2 Application Programming Interface" Revision 2.2.1 2nd May 1997. http://developer.intel.com/ial/winsock2/specs.htm - Intel Winsock Spec WWW page.

[4] D. Comer and R.Yavatkar, Proc. 2nd IEE Symposium on Parallel and Distributed Processing (1990).

[5] http://comet.ctr.columbia.edu/opensig

[6] "A Proposed Flow Specification", IETF RFC 1363, C. Partridge, September 1992.

[7] Resource Management for Distributed Multimedia Applications", Daniel Waddington, Christopher Edwards and David Hutchison, Proceedings of the 2nd

European Conference on Multimedia Applications, Services and Techniques (ECMAST '97), Milan, Italy, 21-23 May 1997. Internal report number MPG-97-10

[8] "The Common Object Request Broker: Architecture and Specification", OMG Doc. 91.12.1, http://www.omg.org/, Dec 1995.

[9] "A CORBA Compliant Real-Time Multimedia Platform for Broadband Networks", Geoff Coulson and Daniel Waddington, Proceedings of the International Workshop on Trends in Distributed Systems (TreDS), Lecture Notes in Computer Science Volume 1161, Aachen, Germany, October 1996. Internal report number MPG-96-33.

[10] "IP Switching: The Intelligence of Routing, The performance of Switching", Ipsilon, February 1996.

Evaluation of Different Video Encryption Methods for a Secure Multimedia Conferencing Gateway

Thomas Kunkelmann [1]   Rolf Reinema [2]   Ralf Steinmetz [1][2]   Thomas Blecher [1]

[1] Darmstadt University of Technology, 64283 Darmstadt, Germany
[2] German National Research Center for Information Technology, 64293 Darmstadt, Germany

Abstract

In multimedia conferencing systems the need for confidentiality and privacy gains more and more in importance, particularly in open networks like the Internet. In this paper we give an overview of the security requirements of multimedia conferencing systems and of applicable security functions. For real-time video transmissions there is a special need for selective encryption of the transmitted data. Existing methods are investigated and their strengths and weaknesses will be shown.

To combine different security functionalities we present the implementation of a scalable security gateway, adaptive to the requirements of specific applications and the properties of special forms of multimedia data.

Keywords: Multimedia communication, Security, Encryption, Partial encryption

1 Introduction

Communication and cooperation in heterogeneous distributed environments are playing a rapidly increasing role in the business processes of today's enterprises. Nowadays several enterprises with distributed locations shift their personal communication and meetings more and more to so-called virtual meetings via computer links. In these cases confidential information has often to be passed securely over open networks like the Internet. Different advances were made to apply security techniques to these forms of communication [BHS94, HJRW96].

Multimedia conferencing supports synchronous communication and cooperation between two or more distributed participants. Characteristic for these systems is the combination of live media like real-time audio and video between different participants and the possibility of sharing documents and applications.

Another kind of distributed multimedia applications with a high demand for security mechanisms are video databases and *video-on-demand* (VoD) services. The security policy for these applications is not focused on optimal protection of highly confidential data, rather on protecting data against illegal access. Therefore the encryption methods needed here tend to be fast, with respect of the high data bandwidth of video streams, and to be cheap to implement in order to supply an emerging market of private users (Pay-TV, VoD). The expense to break into an encryption scheme needs not to be high, but it should be more expensive than the legal access to the video service.

In all these distributed multimedia applications, the cryptographic functionalities must cover different aspects of security, like confidentiality, integrity and authenticity. Therefore different modules of encryption mechanisms must be available to the application. Scalability for encryption methods can be achieved by partial encryption of multimedia data. An elegant way to combine several encryption modules is a *scal-*

able security gateway, providing different security functionalities, adaptive to the requirements of specific applications and the properties of special forms of multimedia data. The main focus of this security gateway considers confidentiality aspects in its current state, but it is extensible for other security functionalities.

The rest of this paper is organized as follows: Section 2 points out the different qualities of security mechanisms needed for multimedia conferencing systems. In Section 3 we present different approaches of partial encryption methods for video streams, since these data need a special consideration due to their huge amount of data and the heavy CPU load they produce. Section 4 presents some methods for the partial encryption of video data streams, as suggested by others and ourselves. Those methods will be evaluated in Section 5. Based on the results found we have implemented a scalable secure multimedia conferencing gateway. Its implementation will be shown in Section 6. Section 7 concludes this paper, giving an outlook on our future work in this area.

2 Security Considerations for Multimedia Conferencing Systems

Nowadays multimedia conferencing systems are widely spread. Multimedia conferencing systems are supporting the synchronous (tele-)cooperation between two or more distributed individuals or groups. In multimedia conferencing, shared workspaces or shared editors are combined with live media like audio and video, which support real-time communication between several distributed sites.

Multimedia conferencing systems can be divided into two main groups:

- centralized or server-based systems, which operate mainly on ISDN and are mostly based on ITU standards (e.g., ITU-T H.320 compliant systems like *Intel ProShare, PictureTel*)
- decentralized or serverless systems, which operate mainly on the Internet and are mostly based on IETF standards (e.g., the MBone-Tools [Eri94], like *vic, vat, wb*)

At the moment, security plays only a secondary role in the development of multimedia conferencing systems. But the need for confidentiality and privacy gains more and more in importance. This applies especially for multimedia conferencing systems that operate on networks driven by third parties or open networks like the Internet.

2.1 Security Requirements

Like any other distributed multimedia system, multimedia conferencing has the following security requirements:

- *access control* to prevent unauthorized access to a conference
- *authentication* to confirm the identities of the communicating partners
- *data confidentiality* to protect data against bugging and to provide traffic flow confidentiality
- *data integrity* to protect data against loss and manipulation
- *non-repudiation* to provide proof of origin and delivery of data

The basic building blocks meeting those requirements are *encryption, authentication, certification* and *integrity preservation*. Relevant methods to fulfill these security requirements are as follows:

- Secret Key Encryption (most commonly used methods: *DES* and *IDEA*, both operating on blocks of 64 bits length)

- Public Key Encryption (e.g. *RSA* and *Diffie-Hellman*)
- Hybrid Encryption (combination of the above two methods)
- Consistency Checking
- Digital Signatures
- Copyright Information (e.g. by digital watermarks)

A general survey of cryptographic methods is given in [Sch96]. The main focus of this paper considers *encryption methods* to provide confidentiality, since their application to multimedia data streams will cause time-critical problems when encrypting the whole data stream. Besides integrity checks, the other security functionalities do not cause any problems concerning the real-time constraints.

2.2 Integration of Security Functionalities in the System

Security functionalities can be built up on two different layers:

- Security in the transmission or networking layer, i.e., security is already provided by the networking protocol used (e.g., SSL, RTP [SCFV96]). An additional data manipulation by security applications is not necessary. For secure communication over ATM networks an approach with a cryptographic protocol unit is presented in [SHB95], able to cope with different encryption keys.
- Security in the data layer, i.e., before data is transmitted from a sender to a receiver it will be manipulated by the appropriate security functions in the application. The security functionality can either be applied to the application as in [BHS94] for a remote conferencing tool, or the application itself is designed to gain security for other programs, e.g. the Secure Shell (SSH) [Ylo95].

One of the drawbacks of network layer security mechanisms is the need for secure underlying transport protocols, which are not available at the moment. IPnG and ATM will provide these functionalities in the near future. The advantage of data layer security is that the transmitted data can be subdivided into parts with sensitive and insensitive data with respect to the human perception.

In comparison to providing security on the data layer, all transmitted data are protected in the network layer. This tends to problems in transmitting huge amounts of data, which is typical for multimedia conferencing. The network layer is not capable of subdividing the data stream in parts with a higher need for protection and parts with a lower or no need for protection. Implementing security functions in the data layer has the advantage that only some parts of the data need to be protected and so the amount of time spent on protecting them can be extensively reduced.

3 Multimedia Data Considerations

Several multimedia data formats require a special treatment in terms of encryption. In particular, these are data formats with real-time properties, like audio and video communication. Here encryption methods cannot be applied straightforward due to the severe time constraints for data processing and the complexity of secure encryption standards. Either encryption must be realized with special hardware, which is not available on many platforms, or the data streams have to be subdivided in order to separate data portions relevant to the human perception for encryption. The latter case is known as *partial encryption* schemes. In this section we investigate this topic in order to examine practicable solutions for scalable encryption mechanisms to be used in our security gateway.

3.1 Data Formats for Video Transmission

For the partial encryption of multimedia data it is important to see how video data is organized in the data stream, in order to develop applicable methods for extracting the relevant data portions. So we first give a short survey over the common data formats used in toady's video conferencing systems. A more general survey can be found in [Ste94] and [StNa95].

Motion-JPEG

The *Motion-JPEG* (M-JPEG) video format is not standardized, it consists of a sequence of single video images (frames) encoded with the *JPEG* format. JPEG (Joint Pictures Expert Group) is a format to encode still images with continuous colors or greyscale values, like natural objects. The JPEG format became an ISO International Standard [JPEG93]. The JPEG image encoding technique leads to a high compression ratio for continuous-toned images. It is based on a combination of applying the *Discrete Cosine Transformation* (DCT) to blocks of 8×8 image pixels, followed by an entropy coding (Huffman and run-length encoding) of the resulting coefficients [Ste94]. The M-JPEG video format is used mainly for video conferencing tools due to a symmetrical expense for encoding and decoding, which is important for real-time applications.

MPEG-1 and MPEG-2

The *MPEG* format for coding and transmitting video signals along with the corresponding audio information has been standardized by the ISO [MPEG93]. For MPEG there are three different standards specified, MPEG-1, MPEG-2 and MPEG-4 (standard scheduled for November 1998). MPEG-1 is today's commonly used video compression standard due to its availability for many platforms and appropriate hardware support. It covers data rates of about 1.2 to 1.85 Mbit/s. MPEG-2 is an enhancement over the possibilities of MPEG-1 by coping high-definition TV and multiple audio channels. Nowadays the first hardware and software solutions for MPEG-2 encoders emerge.

An MPEG data stream is formed of different layers, responsible for the synchronization of audio and video, and providing pre-defined starting points for re-synchronization. MPEG utilizes the compression techniques of JPEG, along with inter-frame relationships (*prediction* and *motion compensation*). The enhancements of MPEG-2 are an extended parameter set for image resolution, pixel aspect ratio and motion compensation techniques. The most remarkable extension is the scalability of video image resolution, leading to an excellent fit into the priority concept of ATM cell transmission, thus making the compression standard interesting for video transmission over ATM links.

H.261 and H.263

H.261 and H.263 are widespread standards adopted by the ITU [ITU96] for transmitting video data streams. The intention of H.261 is to provide video information at a data rate of px64 Kbit/s (with $p \in \{1, ... 30\}$), matching the ISDN specification. Therefore H.261 is toady's mostly used video compression standard for ISDN video conferencing systems (also known as ITU-T H.320 compliant systems) like *Intel ProShare* or *PictureTel*.

The codec (encoding and *decoding* functionality) is designed for a symmetrical encoding and decoding process with a maximum end-to-end delay of 150 ms.

The H.261 standard also specifies many format parameters. The resolutions supported by H.261 are CIF (*Common Interface Format*, 352×288 pixels) and QCIF (1/4 CIF). The frame rate is defined as 29.97 fps. The encoding schemes for H.261 are similar to those used in MPEG, *intraframe* and *interframe* blocks combined with a motion vector perform the basic units of an H.261 image.

H.263 enlarges H.261 by providing more resolution formats, better prediction methods for motion compensation, better error correction schemes and higher compression ratios than H.261. H.263 is not restricted to data rates of p×64 Kbit/s anymore. Because of these enhancements it will probably replace H.261 in the near future.

3.2 Performance Aspects for Encrypted Video

As pointed out in [BGU95], modern high-performance workstations and servers are capable of playing MPEG-1 or M-JPEG video, leaving about 20 to 60 percent CPU time for other jobs when using hardware JPEG support. On most desktop workstations such a computing power is not available. Here the frame rate or the pixel resolution has to be reduced (e.g. from CIF to QCIF format) to meet the limited CPU capacity. Performance measurements on a PC (100 MHz Pentium, Linux) showed that such a system can playback about three H.261 QCIF video streams with frame rates sufficient for video conferencing (between 11 and 12 fps).

Table 1 shows the performance evaluations of several hardware platforms decrypting video streams in software, with standard library implementations of the DES algorithm. The reason for investigating DES is the fact that cryptanalysts consider it to be a safe algorithm for ciphertext-only and known-plaintext attacks [Sch96], except for the key length of 56 bits, which opens a door for brute-force attacks with massive hardware power. The IDEA algorithm [Lai92] is an alternative which provides a far better resistance against such attacks by using a key length of 128 bit, its implementations are slightly faster as DES [Sch96]. For modern encryption methods, e.g., FEAL and Blowfish [Sch96], this safety cannot be guaranteed due to the short time period in which cryptanalysts could gain experiences with them. They may become a faster alternative if time will show that there exist no known attacks against them.

DES CPU usage	1.5Mbit MPEG	2Mbit M-JPEG	3×128 Kbit H.261
Intel Pentium-100, Linux	86.70 %	★115.62 %	21.67 %
DEC Alpha 1000/ 266	65.63 %	87.50 %	16.41 %
Sparc 20 (Solaris)	76.01 %	★101.34 %	19.00 %
Sparc 4c (SunOS)	★312.77 %	★417.03 %	78.19 %

Table 1: CPU utilization of different hardware systems for DES software encryption. The MPEG and M-JPEG cases represent e.g. Pay-TV scenarios (16 - 25 fps), while the H.261 scenario describes an ISDN video conference with three video channels open (12 - 15 fps).

For the MPEG and M-JPEG scenarios we examined, the need for reducing the encryption effort is obvious, the slower workstations are already overloaded with the DES decryption (★ = projected values). For the H.261 scenario, an encryption CPU usage of 20 percent implies a frame reduction from e.g. 11 to 8.8, violating the lower bounds for human image perception. Therefore partial encryption is a suitable solution also for this case, as well as the usage of one of the modern, faster encryption algorithms, with the drawback of no guarantees for the security of these algorithms and maybe no availability of the decryption algorithm at the receiver's site.

Even if there will be more powerful processors in the future, there is no need to waste their whole CPU time for video decryption.

4 Partial Video Cryption Methods

Considering the results from performance measures in secure video systems, several methods for partial encryption of video data have been proposed in the last few years, which are summarized in this section.

4.1 SEC-MPEG

SEC-MPEG [MeGa95] is a toolkit for partial encryption of MPEG-1 data. The development of SEC-MPEG is based on the Berkeley-MPEG player [PSR93]. The aim of this toolkit is to achieve confidentiality and integrity checks. Confidentiality is achieved by using the DES algorithm, integrity checks are carried out by a *cyclic-redundancy check* (CRC) due to performance issues, at the expense of a weak integrity certification. The toolkit supports four levels of confidentiality (*C-levels*) and three levels of integrity (*I-levels*), beginning with encrypting the header information, up to an encoding of the whole MPEG stream. In C-level 2 a subset of DCT blocks is selected, which will be partially encrypted, while C-level 3 encrypts all intracoded image information.

4.2 Partial encryption of intracoded frames

Some work has been done in partially encrypting only the intracoded frames (I-Frames) of an MPEG stream [MaSp95] or the intracoded blocks in intercoded frames, as in C-level 3 of SEC-MPEG. In [AgGo96] an example of this kind of encryption is given, the authors also show the limits of this technique (which is also used in SEC-MPEG). Video sequences with a high degree of motion still show a lot of details of the original scene. As a remedy the increase of intracoded-only frames is suggested, but this will also vastly increase the size of video data. In the case of video transmissions over a channel with limited bandwidth this can be no solution.

4.3 Encryption of DCT block information

A method for an encoding/ decoding process with no significant delay resulting from additional encryption is applicable to video compression techniques based on the JPEG algorithm. In [Tan96] this method is described for the MPEG standard. It is based on the zigzag ordering of the DCT coefficients before entropy coding is applied. This order is randomly permuted, the secret key of the encryption is the permutation itself. Due to a table lookup for computing the zigzag coefficients in usual video decoders this permutation generates no temporal overhead. The drawback of this method is the worse performance of the run length encoding of the DCT coefficients, which results in an expansion of the encoded video data of about 20% to 40% for the tested video sequences.

4.4 Reducing the amount for strong encryption

Statistical analysis of MPEG streams show that it is still sufficient to reduce the effort for encryption to one half of the video stream, and use these data as a *one-time pad* for the other half of the stream, in order to obtain a strong cryptographic protection for the whole MPEG data [QiNa97]. The proposed algorithm parses the MPEG stream down to the slice layer, but does not touch the macroblock information. So it is suited to operate in a separate encryption module, independent of the encoding or decoding process. The method needs about 53% of the effort for encrypting the whole data stream, its drawbacks are the usage of multiple encryption keys and the overwriting of some MPEG header fields, which makes the solution infeasible for most existing applications.

4.5 Scalable method for JPEG-based video

In [KuRe97] we present a scalable partial encryption method, which allows a security level of nearly every granularity. It can be applied to all video compression methods based on the JPEG standard, in particular the formats mentioned above. This method is not prone to the motion prediction problems mentioned in 4.2. Our method takes advantage of decreasing importance for the image composition of the DCT coefficients, so it is sufficient to encrypt only the first few of them. The algorithm starts with encrypting a data block at the beginning of a DCT block and guarantees the protection of at least the first n DCT coefficients of a block, encrypting consecutive data portions in the video stream of the encryption method's block size. The parameter n of encrypted coefficients provides scalability for the security level. Figure 1 gives an example (with $n=3$), which parts of an MPEG streams will be encrypted.

Figure 1: Encrypted parts of a video stream with our partial encryption method

4.6 Combination of Bitstream- and VLC Encryption

We performed some experiments on the partial encryption of "confidential" video material to find practicable solutions with low encryption effort, but a high level of confidence for the protected video stream. The experiments were made with a modified version of the Berkeley MPEG encoder and player [PSR93]. We have investigated three approaches for encryption:

- Encryption of the *DCT coefficients* before they are encoded with the Huffman tables of MPEG or H.263.
- Partial encryption of the *Variable Length Codes* (VLC), which occur after applying the Huffman encoding.
- Selective encryption of the MPEG or H.263 data stream, depending on the index of the coefficients represented by the VLC.

The first approach is only useful for security analysis and the computation of optimal parameters for the other approaches. Due to a pseudo-random bit distribution after encryption, the DCT coefficients cannot be compressed effectively, resulting in video sequences which occupy the same disk space as the raw digitized material.

Our experiments have shown that a combination of VLC encryption and bit stream encryption is most useful for partial encryption schemes, when keeping a high level of confidentiality is the primary goal.

4.7 DVB - Conditional Access

Conditional Access (CA) is a method for video encryption used in the *Digital Video Broadcasting* (DVB) project [DVB96]. It provides a *Common Scrambling Interface*, which is supported by every DVB program vendor. It is a combination of a block cipher and a stream cipher and is fed with two control words, which are transmitted in the video control stream. The individual access to specific programs of a video transmission (pay-per-view) is regulated by *Entitlement Management Messages* (EMMs), which are only valid in combination with a unique receiver ID number.

The system supports the encryption of a whole MPEG-2 transport stream, or the encryption of specific packeted elementary streams, e.g. only a video or an audio stream. The encryption unit can switch between two keys, so a key exchange is possible during a video transmission.

5 Evaluation of Results

We first present some aspects on the safety of partial encryption methods for video data. Based on these considerations, we compare the different methods with respect to safety, time consumption and communication overhead.

5.1 Possible Reconstruction of Protected Data

With methods used in cryptanalysis, e.g., statistical and entropy evaluations, it may always be possible to detect those portions of a data stream which have been encrypted. However, this will be a difficult job for partially encrypted (MPEG or similar encoded) video streams due to the nearly redundancy-free Huffman encoding. An eavesdropper who succeeded in analyzing a partially encrypted video stream might probably reconstruct a video frame as in the examples of Figure 2. Here the non-reconstructible protected information is set to zero, otherwise the random encrypted information would still obscure the reasonable information.

These examples motivate to protect truly confidential video information with an adequate method, e.g. the scalable approach presented in [KuRe97]. In other scenarios, where encryption is merely used to aggravate the access for the public, e.g., video-on-demand systems, the expense for reconstructing parts of a video is out of all proportion to the fee for joining the movie broadcast legally. In these scenarios a simple encryption method might be considered as sufficient.

Figure 2: Maximal possible reconstruction for intracoded block encryption (left) and with the method of [KuRe97] (right), both frames with about 46% encrypted data (video *flowers*, 1/2 original size, the original frame image is shown in Figure 7).

5.2 Experimental Results

Our experiments are based on a series of different video sequences, which reflect several scenarios where digital video can be used. Movies for video-on-demand (VoD) applications are represented by the test sequences "Flowers" and "Biker" (action movie), several sporting scenes ("Soccer" and "Skating") with different ratio of movement, and video conference scenes ("Conference", portrait of a male speaker, and "Talk", two talking persons with the fade-in of a telephone number) are used for testing.

In VoD scenarios the encryption effort need not to be high, even with a few percent of encrypted data the quality of the video material becomes intolerably poor. In Figure 3 we present an example for an encrypted video image with about 25% of the data encrypted. We consider about 10 percent encryption as a satisfactory level for VoD applications, which complies with the fact that here the software and hardware effort must be minimized to keep the costs per set-top unit cheap.

For truly confidential video sequences (e.g. the phone number in the "Talk" sequence) it is not sufficient at all to pick some few video blocks or DCT coefficients for encryption, as it is done in most partial encryption schemes. Here the combination of bit stream and VLC encryption seems a good approach, although the security of the stream cipher method in our VLC encryption is considered not to be as secure as, e.g., DES. A replacement for this algorithm with a block cipher would provide a security benefit. When using our scalable approach it is necessary to protect at least the first 10 to 12 DCT coefficients in order to keep a high level of confidence. This results in an encryption rate of 40% or more of the video data. To protect numbers or letters in the video image from being read by an eavesdropper, a ratio of 50% encrypted data is necessary.

Figure 3: Video sequence *biker* with 25% encrypted data, playback (left) and maximal possible reconstruction (right)

5.3 Comparison of the Encryption Methods

In Table 2 we compare the different partial encryption methods with respect to security, scalability, time effort, protocol signaling overhead and feasibility for the usage in a security gateway for video conference applications.

Method	Security	Scalability	Time overhead	Protocol overhead	Isolation
SEC-MPEG	high	3 levels	DES encryption	about 17 to 32%(own data format)	possible
Frame-type encryption	high	I: 25-40% IP: 70-85% IPB: 99%	DES encryption	none	possible (low overhead)
Intra-block encryption	high	no	DES encryption	none	possible
DCT permutation	breakable	no	none	none, data volume + 20% to 40%	not possible
Scalable method	high	full, from 8% to 100%	DES encryption	3-5%	possible (10.5%)
VLC/ bit stream combination	high (medium for VLC bits)	full	DES encryption + VLC re-order	3-5%	possible
DVB - Conditional Access	n.a. (details are secret)	no	Hardware: none, SW: low (XOR)	yes (control words, EMM, etc.)	default

Table 2: Comparison of different partial encryption methods

One important aspect for the implementation of a partial encryption method in a security gateway, independent form the video coding unit, is the possibility of separating the data portions relevant for encryption in an efficient fashion (*Isolation* in Table 2). For a method like e.g. [Tan96], this implies a complete decoding and re-encoding of the whole video stream, which is not feasible. For the scalable method of [KuRe97] the gateway process must identify the start of a DCT block in the video sequence. In [PSR93] the parsing time is estimated with 17.4% of the whole play-back time, [MeGa95] achieved values of about 30%. Our experiments showed, however, that this time can easily been reduced to a value of 10.5% of the total playback time for the complete separation of encryption data. Re-implementing the parsing algorithm with a finite state machine might furthermore reduce this time effort. So this partial encryption algorithm is well suited for an operation in a separate security gateway.

An integrated solution of the decoding process and the security gateway needs only about 1.01% time overhead for the separation of encrypted data, again compared to the video playback time.

Another important factor is the signaling or control data overhead an encryption scheme generates. These data can be embedded in the video stream as it is done in SEC-MPEG with a special encryption header flag, or it can be transmitted via a separate control channel as in our security gateway approach. Using a special encryption header flag implies a length of at least 32 bit, resulting in a protocol overhead of 17% (intercoded) to 32% (intracoded) additional bandwidth. In our solution the signaling data is encoded with fewer bits, so the total overhead is about 4.1% in average, at a level of ten coefficients protected.

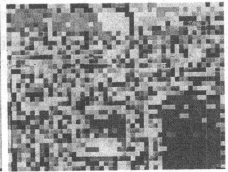

Figure 4: Video clip "skating" with encryption in ECB mode. The uniform areas of the original image are also visible in the encrypted frame

The impact of the encryption method used can be seen in Figure 4, where DES in standard ECB mode (Electronic Codebook, the same input pattern generates identical encryption output) was chosen. Large uniform areas in the video image result in identical block encoding and can be identified in the encrypted output data. Using CBC mode (Cipher Block Chaining, the output data is fed back to the next encryption block) avoids this security leak. This method has the slight drawback that any bit error results in the fact that the following data block also cannot be decrypted.

Figure 5 shows the difference between VLC encryption and encryption of only some DCT values. Fine-scaled objects like the text across the video image are expressed by higher DCT coefficients and thus can be reconstructed from a partially encrypted video. So we suggest also to protect the VLC coefficients to overcome this security leak.

Table 3 shows that there is always a gain in the frame rate when using partial encryption in software solutions. The performance measures were calculated on an UltraSparc station with Solaris. For common desktop PCs with no multimedia support the performance gain will be comparable. In the case of inter-coded frames, the encryption rate was 35% (VoD example), for the intra-coded (Motion-JPEG, video-conference scenario) frame example, the encryption rate was assessed with 50%. The experiments show that the performance gain is at least two frames/ second, with an effective frame rate of about 12 to 15 fps this results in a visible profit.

Video clip	Mode	No enc.	Full enc.	Partial enc.	Encryption rate
Skating	intra	20.82	15.14	17.92	50%
Talk	intra	17.86	12.70	15.28	50%
Soccer	intra	16.94	10.88	14.02	50%
Skating	inter	25.50	22.52	24.32	35%

Table 3: Frame rates(fps) with full and partial encryption

Figure 5: Video clip "talk" original (top left), 1 DCT coefficient protected (right), 1 VLC bit encrypted (bottom left), combination of both (right)

6 Implementation of a Scalable Secure Conferencing Gateway

For the secure end-to-end transmission of multimedia conferencing data streams over open internetworks between two or more distributed sites an application independent *secure conferencing gateway* has been implemented. It provides the following advantages:

- security can be achieved even if the used conferencing applications do not support it
- conferencing systems can be implemented independently of the used encryption methods

Currently the secure conferencing gateway is only restricted to the secure transmission of video data streams using the above described methods. It provides various ports for the different applicable video compression modes: M-JPEG, MPEG-1/ MPEG-2 and H.261/ H.263.

The gateway consists of two parts, an encryptor and a decryptor using our scalable encryption method. Between two or more such gateways there are two channels established, a data channel that carries the encrypted video data and a control channel. The control channel is used for authentication, exchange of session keys, the security level and synchronization during the session (e.g., changing the session key or the security level during a transmission). This architecture is characterized by its high degree of modularity and scalability to reflect different application scenarios and different levels of security needed in various conferencing situations.

Unlike in [BHS94] where all session keys are transmitted to the participants in advance by secure e-mail, the session keys were chosen by random during the conference. The initial session key is chosen by the first site entering a conference. When the next site enters the conference, it recognizes that there is already another participant and requests the current session key from it. The session key is then encrypted using the public key of the requester. The initial security level is set to a default value, which is suitable for typical conferencing situations.

The session key and the security level can be changed by any site at any time, which requires synchronization through the control channel. First, the change of the session key or security level will be announced. After the acknowledgment of all sites, changes take place at the pre-scheduled time.

The implementation of the secure conferencing gateway is not just restricted to point-to-point communications between two distributed participants. It can also be used for point-to-multipoint transmissions between two or more distributed sites. To achieve this goal we make use of IP-Multicast [Eri94].

In some application scenarios like seminars and discussions people often enter and leave a conference at their convenience. This is reflected in the fact that there is no master or server gateway that would supervise the change of session keys, security levels and ensure the consistency between all involved gateways. The change of the master role would generate more communication overhead than the distributed negotiation for changing the session keys and security levels using IP-Multicast.

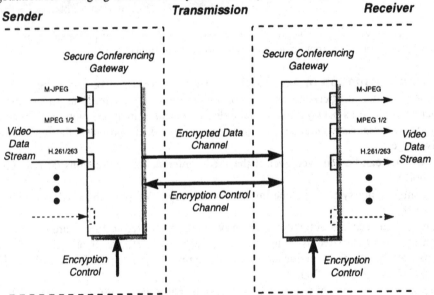

Figure 6: Secure Conferencing Gateway

The implementation of the partial encryption algorithm is based on the Berkeley MPEG encoder/ decoder [PSR93]. To provide encryption, decryption and certification of public keys SecuDE is being used [GMD97]. SecuDE is a toolkit (for UNIX and MS-DOS/MS-Windows platforms) which offers a library of various security functions. It provides basic cryptographic functions (like RSA or DES), digital signatures, X.509 key certification, operation of certification authorities, secure access to public X.500 directories for the storage and retrieval of certificates, cross-certificates and revocation lists. SecuDE provides a so-called Personal Security Environment (PSE) which contains the user's private and public key pair.

7 Summary and Future Work

In this paper we pointed out the different security requirements needed for multimedia conferencing systems. A special treatment has to be applied for live media, especially for real-time video data due to the large amount of data to be protected. Partial encryption is a solution to solve this problem.

MPEG-1/MPEG-2 and H.261/H.263 are widespread compression standards used in most of today's video conferencing applications. They are well suited for partial encryption because on the one hand they make use of DCT, which has a high potential for dividing data in more relevant less relevant parts (entropy of the coefficients). On the other hand, large amounts of video data are encoded by reference to preceding or following blocks (intracoded blocks), from this it follows that only the referenced blocks have to be protected. Motion-JPEG, which also uses DCT, is not well suited for all partial encryption schemes. In contrast to MPEG-1/MPEG-2 and H.261/H.263 it does not make use of referenced frames or blocks and so the redundancy of Motion-JPEG data streams is much higher.

The newly emerging *MPEG-4* standard has to be further examined with respect to encryption. MPEG-4, which is still in the specification phase, provides specific solutions for different types of objects during the encoding process because of this it might offer totally new possibilities for partial encryption methods.

The implementation of our scalable secure conferencing gateway is based on the performed evaluation. One of the next steps in the gateway implementation is the integration of security functions for other media beyond video data streams.

With the specification of the new *Security Service* as a *Common Object Service* for CORBA (Common Object Request Broker Architecture) by OMG (Object Management Group) end of last year [OMG96], the integration of our security gateway in a CORBA security service will be investigated. Along with the integration in a middleware platform for heterogeneous environments, a new service called *transcoding* [AMZ95] can additionally be offered. Transcoding means the translation between different encoding formats, transmission protocols and bit rate adaptation. The specification in CORBA IDL will make it much easier for application programmers to use these services particularly in open distributed environments, since the service interfaces are then accessible in a standardized manner.

Literature

[AgGo96] I. Agi, L. Gong: *An Empirical Study of Secure MPEG Video Transmissions.* ISOC Symposium on Network and Distributed System Security, San Diego, CA, 1996

[AMZ95] E. Amir, S. McCanne, H. Zhang: *An Application Level Video Gateway.* Proc. 3rd ACM Multimedia, San Francisco, CA, 1995

[BGU95] P. Bahl, P.S. Gauthier, R.A. Ulichney: *Software-only Compression, Rendering, and Playback of Digital Video.* Digital Technical Journal Vol. 7(4), 1995

[BHS94] K. Bahr, E. Hinsch, G. Schulze: *Incorporating Security Functions in Multimedia Conferencing Applications in the Context of the MICE Project.* Proc. 2nd Int'l Workshop IWACA'94, Heidelberg, Germany, 1994

[DVB96]   *The Digital Video Broadcasting Project.* http://www.ebu.ch/dvb_home.html, 1996

[Eri94]   H. Eriksson: *The Multicast Backbone.* Comm. of the ACM, 37(8), pp. 54-60, 1994

[GMD97]   W. Schneider (Ed.): *SecuDE Documentation.* ftp://ftp.darmstadt.gmd.de/secude/, Darmstadt, Germany, 1997

[HJRW96] E. Hinsch, A. Jaegemann, I. Roper, L. Wang: *The Secure Conferencing User Agent - A Tool to Provide Secure Conferencing with MBONE Multimedia Conferencing Applications.* Proc. IDMS '96, Berlin, Germany, 1996

[ITU96]   ITU-T Recommendation H.263: *Video coding for low bit rate communication.* 1996

[JPEG93] ISO/ IEC International Standard 10918: *Digital Compression and Coding of Continuous-Tone Still Images.* 1993

[KuRe97] T. Kunkelmann, R. Reinema: *A Scalable Security Architecture for Multimedia Communication Standards.* Proc. 4th IEEE Int'l Conference on Multimedia Computing and Systems, Ottawa, Canada, 1997

[Lai92]   X. Lai: *On the Design and Security of Block Ciphers.* ETH Series in Information Processing, 1, H. Gorre Verlag, Konstanz, 1992

[MaSp95] T.B. Maples, G.A. Spanos: *Performance Study of a Selective Encryption Scheme for the Security of Networked Real-time Video.* Proc. 4th Int'l Conference on Computer and Communications, Las Vegas, NV, 1995

[MeGa95] J. Meyer, F. Gadegast: *Security Mechanisms for Multimedia Data with the Example MPEG-1 Video.* http://www.cs.tu-berlin.de/~phade/secmpeg.html, 1995

[MPEG93] ISO/ IEC International Standard 11172: *Coding of Moving Pictures and Associated Audio for Digital Storage Media up to about 1.5 Mbit/s.* 1993

[OMG96]   Object Management Group: *Security Service: v1.0.* 1996

[PSR93]   K. Patel, B.C. Smith, L.A. Rowe: *Performance of a Software MPEG Video Decoder.* Proc. ACM Multimedia, Anaheim, CA, 1993

[QiNa97] L. Qiao, K. Nahrstedt: *A New Algorithm for MPEG Video Encryption.* Proc. 1st Int'l Conf. on Imaging Science, Systems and Technology, Las Vegas, NV, 1997

[SCFV96] H. Schulzrinne, S. Casner, R. Frederick, V. Jacobson: *RTP: A Transport Protocol for Real-Time Applications.* RFC 1889, 1996

[Sch96]   B. Schneier: *Applied Cryptography.* 2nd Edition, ISBN 0-471-11709-9, John Wiley, New York, 1996

[SHB95]   D. Stevenson, N. Hillery, G. Byrd: *Secure Communications in ATM Networks.* Comm. of the ACM, 38(2), pp. 45-52, 1995

[Ste94]   R. Steinmetz: *Data Compression in Multimedia computing - standards and systems.* Multimedia Systems, 1(4), pp. 187-204, Springer Verlag, Berlin 1994

[StNa95] R. Steinmetz, K. Nahrstedt: *Multimedia: Computing, Communications and Applications.* Prentice Hall, München 1995

[Tan96]   L. Tang: *Methods for Encrypting and Decrypting MPEG Video Data Efficiently.* Proc. 4th ACM Int'l Multimedia Conference, Boston, MA, 1996

[Ylo95]   T. Ylönen: *The SSH (Secure Shell) Remote Login Protocol.* http://www.cs.hut.fi/ssh/RFC, 1995

Figure 7: Original video images of Figure 2 and Figure 3

The FreeBSD Audio Driver

Luigi Rizzo

Dip. di Ingegneria dell'Informazione
Università di Pisa
via Diotisalvi 2 – 56126 Pisa (Italy)
email: l.rizzo@iet.unipi.it

Abstract. We recently developed an audio driver in the FreeBSD operating system. In this work, we decided to consider compatibility with existing software interfaces only as a secondary issue, to be implemented at a later time and only for those applications which could not be adapted to the new software interface. This turned out to be a significant advantage, since it let us design the driver (and particularly, its software interface) looking at the real needs of applications, rather than duplicating existing, old interfaces, and having applications adapt (in many cases suboptimally) to what the driver could offer.

The main results of our work is the definition of a software interface for audio devices which is well suited to multimedia applications. The new interface is small, simple but powerful, and allowed several simplifications, and significant performance enhancements, in the applications. In this paper we motivate our design choices, illustrate our interface, and discuss implementation issues both for the device driver and applications. The software described in this paper, and appropriate application routines, are available from the author.

Keywords: Multimedia, audio conferencing, audio devices, operating systems.

1 Introduction

Networked multimedia applications have become very popular in recent years [4, 6, 12, 13], due to a number of enabling factors such as the availability of high performance computing hardware with multimedia capabilities, which have made it possible to run in software powerful compression/decompression algorithms, and advances in network and modem technology, which have made network connectivity very widespread and with suitable data rates.

Among multimedia applications, the most compelling ones are those requiring (soft) real-time features, synchronization among streams of different types, or full duplex operations. Audio/video conferencing systems are typical examples where these demanding requirements appear.

Operating system support for multimedia devices, however, predates these applications. Audio CODECS and video grabbers are conceptually simple devices, and in principle they can be easily integrated into an operating system (e.g. Unix) using standard primitives such as `read()` and `write()`, plus a handful of `ioctl()` calls to access special device features. Such a primitive audio device driver[1] is what is generally available in Unix and other systems, under the assumption that additional functionalities can be supplied by library routines, or by passing audio data through a server process which acts similarly to the X Window server (see Fig. 2).

The reason why the above reasoning fails is that the server will ultimately need the services of the device driver as well. As more and more functionalities are put in the hardware (e.g. by offloading operations to a DSP in the audio card) the software interface of the device driver will need to be extended to effectively support these functionalities.

But what are really the needs of applications which require support within the device driver ? Buffering is often almost all what is needed if one of the communicating parties has no strong timing requirements (e.g. it is a disk), and can adapt to the average data rate that the CODEC imposes. However, full duplex applications, or other applications with real-time constraints (e.g. when audio must be synchronized with visualization data), have some timing requirements, since they need to act upon specific events (e.g. when a given sample is acquired or played out). This in turn requires the software interface to the device to support the exchange of synchronization information. When these are not available (e.g. because they cannot be easily accomodated in some preexisting device driver structure, or because they would break backward compatibility), generally the only alternative is to enable non-blocking operation on the audio device, and implement synchronization-related functionalities in other ways (e.g. by means of timers). But this is an expensive solution, since it makes applications busy-wait for events instead of using the synchronization mechanism which are available in the kernel. Yet this is a common approach, which is bearable only because today's workstations are sufficiently overpowered to tolerate even highly inefficient operating systems and applications.

Recently, we were involved in the development of a new audio driver for the FreeBSD operating system. FreeBSD did have a freely available device driver for sound cards, but it lacked support for newer cards, lacked functionalities (e.g. full duplex support) and had some limitations which applications had to circumvent at the price of performance, efficiency or program clarity. Our goal was then to produce a new driver to overcome all these limitations, possibly redefining the software interface of the driver. On the other hand, the existing software interface [10] was sufficiently popular to require support in the new driver.

This gave us two alternatives: either build a driver which was compatible with the existing one, adding new features on top of this; or start the design without the constraint of backward compatibility, and add a compatibility layer at a later time, only for those applications for which source code was not available and adaptation to the new interface was not possible. Clearly the second approach was more demanding, in that it required a study of existing applications to determine which features were needed and how they were used, and some

---

[1] In this paper we will not consider specialized hardware which exists for generating music, e.g. FM or wavetable synthesizers. These devices have a completely different internal structure and software interface.

porting effort to adapt applications to the new interface. However, this approach proved to be very fruitful. In fact, the analysis of applications let us understand what the driver *should really* do, rather than what the driver *used* to do. This analysis, not without surprise, showed that many applications were adapted to the existing audio driver in a quick and dirty way, by making inconsistent use of the available functions, calling functions with the same or similar semantics in mixed or redundant ways, and very often using features (such as non-blocking I/O) which were not necessary or highly expensive. As a result of this analysis, we had a chance of identifying which of the existing functionalities were really needed, which ones were merely duplicates, which ones were probably useless, and which ones were missing. This served us as a guide to what we should really implement in our driver and, at the same time, to how the applications we studied should make use of the audio functions.

Basing on this work, we have defined what we believe to be a simple yet powerful set of functions to access audio devices. These functions can be easily and efficiently implemented in device drivers, let applications make full use of the available synchronization mechanisms, and give very good support for highly demanding multimedia applications. In this paper we provide a detailed description of the functionality and the semantics of our primitives, motivating their existence and showing their usefulness. The software interface described in this paper has been implemented and application have been modified to make use of it, often with large improvements on the efficiency of the applications, and also with some simplifications in the code.

The paper is structured as follows. We start with a basic overview of the hardware which implements audio devices. In Section 3 we describe the methods to access the device to transfer data, and motivate our choices in relation with some existing applications. In Section 4 we discuss the synchronization requirements between applications and the driver, and show how a small set of calls can be used to implement them efficiently. Related work is presented in Section 5. Finally, in Sections 6 we discuss some implementation issues and show how the new features were used to improve the performance of existing applications.

2 Audio hardware

Figure 1 shows the hardware/software structure of a typical audio device. We have two logically independent subsystems[2], one for audio capture and one for audio playout.

The capture section contains a *mixer* to select the input source which is routed to the Analog to Digital Converter (ADC). The ADC output is stored into a circular buffer, generally by means of some DMA engine. The hardware dictates the native resolution (e.g. 8..16 bits), data formats (e.g. linear or companded using μ-law), channels (mono/stereo) and sampling rate, whereas the software

---

[2] except for cases where hardware limitations make the subsystem share resources thus preventing full duplex operation.

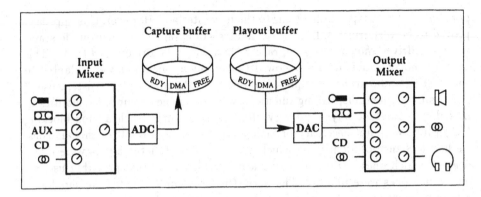

Fig. 1. A model of audio devices. On the left is the capture section, on the right the playout section.

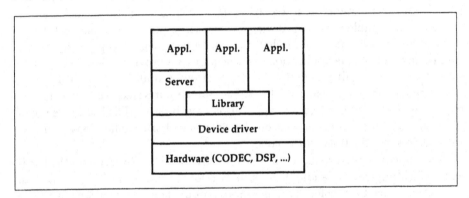

Fig. 2. Possible structure of applications using the audio device. Access can be direct, through a library, or through a server.

implements the circular buffer and may additionally implement some data format conversion before passing data to the application program.

The playout section uses a circular buffer to hold data which are sent to the Digital to Analog Converter (DAC), generally using a DMA engine. The output of the DAC is routed through a mixer, together with other sources, to the appropriate output. Again, the operating system software implements the buffering strategy and possible format conversions to supplement the native features supplied by the hardware.

The DAC and the ADC are usually part of the same physical device called *CODEC*. The transfer of samples between the CODEC and the buffers supplied by the operating system can be done in several ways. Most of the times, the audio device uses a DMA channel which is present on the system's mainboard (as in the case of Intel-based PC's). In some other cases, the audio device has an internal DMA engine which acquires control of the bus and performs the data transfer (devices on the PCI bus generally operate in this way). As an alternative, the audio device could have some internal memory where data are buffered (e.g. under control of a specialized processor), and the main processor has only to transfer blocks of data at given intervals, possibly using programmed I/O.

In some cases the audio device has an onboard Digital Signal Processor (DSP) which can be used to run data compression algorithms, offloading the main processor from this task. This is in general a good idea since some algorithms (e.g. GSM or MPEG encoding/decoding) are expensive and it is much cheaper to run them on a DSP (which has become a commodity device since it is extensively used in cellular phones and other digital audio devices).

There are large variations in the capabilities of the mixers as well. Some simple devices just have one input and one output channel, directly connected to the ADC and the DAC, respectively, with no volume controls. The majority of audio devices for PC's have a simple multiplexer with only a master volume control on the input section, and a full featured mixer with independent volume controls on the various channels on the output section. Newer devices, finally, have full-featured mixers on the input path as well, together with a larger selection of sources and the ability of performing miscellaneous functions such as swapping left and right channels, muting sources, etc.

3   Software interface to the audio device

In this Section we start the presentation of our software interface to audio devices. We focus our attention on the "device driver" layer in Figure 2, and particularly to the transfer of data between the audio device and the application program. Synchronization primitives will be discussed in the next Section.

Prior to using the audio device for data transfer, applications will need to acquire the capabilities of the device (supported sampling rates, data formats, number of channels, full duplex operation, and any other device-specific "feature"[3]), and to set the desired data formats. These operations are usually done by means of some ioctl() calls, whose name and format change from system to system. Although some approaches (e.g. those where all parameters are read or set at once, using a single call) appear to be more elegant and efficient than others, there are in practice no differences since this is generally a one-time operation.

Another one-time operation, for audio devices which include a DSP, might be the downloading of appropriate firmware to the DSP to perform the required function (e.g. running some compression/decompression algorithm). Dedicated software interfaces are generally used for this purpose, and efficiency is generally not a primary concern.

---

[3] There are significant differences among audio cards. As an example, some CODECS can only work in full duplex under some constraints, e.g. using 8-bit format in one direction and 16-bit in the other one. Other CODECS have bugs which are triggered by certain operating modes, and so on. The driver can block erroneous requests, but the only way to make good use of the available hardware is that the driver provides a unique identifier for the actual hardware and applications (or libraries) can map these identifiers to a list of features and adapt to them.

Finally, appropriate ioctl() calls are necessary to control the mixers which are present in the signal paths. This is an area which would really benefit from some standardization effort, given the significant differences in capabilities of these devices. However, the control of mixer devices is conceptually simple since the requested operation (e.g. setting a volume or selecting an input source) generally takes place in real time, and the only significant issue is to have an interface which can ease the porting of software to different systems.

We will not discuss the three items above (selecting data format, programming onboard devices, and controlling the mixer) in further detail in this paper.

3.1   Character versus block mode

Traditionally, audio devices have been used as character devices, with the granularity of access supported by the driver being the single byte. Applications however often transfer data in blocks of fixed size. This is not only a matter of convenience, it is also a need dictated by the use of compression algorithms which operate on blocks of samples with a given size. This pattern of access has implications on the type of operating system support required by such applications.

Multimedia applications usually have to deal with multiple events (e.g. audio input, data coming from the network, GUI events, timers, etc.) and they cannot block on a single source waiting for data to be ready. The usual solution to this problem is to use a blocking primitive (such as the Unix select()) which allows the process to block specifying a set of events which the process is waiting for. If the audio device is treated as a character device, a select() on the device might return as soon as a single byte can be transferred – at most 125 μs with the typical sample rate of audioconferencing tool, or even less for high quality audio. What we would really want, instead, is to specify to the driver the granularity of the select() operation, so that it will not wake up until the desired amount of data is ready[12]. Some software interfaces, e.g. OSS (formerly Voxware) [10], allow this but only indirectly, since they permit to choose among a small number of block sizes for DMA operations.

To fulfill this goal, our audio driver has two modes of operation, *character* and *block* mode. In *character* mode, the device produces a stream of bytes, and select() returns when one byte can be transferred. To enter the *block mode* (not to be confused with *blocking mode*, which is an orthogonal feature), the AIOSSIZE ioctl() is used to specify the granularity to be used for the select() operation. The latter will return successfully only when *at least* a whole block of data can be transferred. AIOSSIZE can modify the requested value if it is not an acceptable one (e.g. the requested block size exceeds the size of the internal buffer) and returns the block size in use. A block size of 0 or 1 will bring the device driver back into character mode.

We want to point out that the AIOSSIZE function only specifies the behaviour of the select() call. In our driver, both read() and write() retain the usual byte granularity. We found this to be a necessary feature because it permits applications to control the data transfer rate with fine granularity, rather than being forced to use multiples of the block size. For robustness reasons, by no means the user should make further assumptions on the behaviour of read() and write(). In particular, it should not be assumed that they always transfer the requested amount of data, or that they always work on multiples of the block size. Similarly, the user should make no assumptions on the internal operation of the driver (e.g. that the granularity of DMA transfers equals the value specified with AIOSSIZE). Such assumptions might make programs not check error conditions (e.g. read() or write() returning with a short count) which might indeed occur depending on the internal implementation of the driver or the features of the hardware.

3.2 Support for non-blocking I/O

Traditionally, device drivers provide support for non blocking I/O. This operating mode can be selected at device open time, or by issuing the FIONBIO ioctl(). In non-blocking mode, read and write operations will never block, at the price of possibly returning a short transfer count[4]. Non blocking reads are possible, even in blocking mode, by invoking the FIONREAD ioctl() first, and then reading no more than the amount of available data.

There is no standard function which is a dual of FIONREAD for write operations. In our driver we have implemented such a function, called AIONWRITE[5], which returns the free space in the playout buffer. A write of this many bytes will not block, even if blocking mode is selected. In our implementation, both FIONREAD and AIONWRITE track closely the status of a DMA transfer.

4 Synchronization

In this Section we discuss in more detail those which we consider useful features to support synchronization of audio with other activities.

It is important for some applications to know the exact status of the internal buffers in the device driver, both in terms of ready and free space, because some other activities should be synchronized with the audio streams. As an example, a player program wants to avoid that the playout buffer becomes empty in the middle of operation; or, a telephone-like application might need to check that the buffers do not become too full, in order to control the end-to-end delay.

When data transfer occurs at a constant nominal rate, in principle this information can be derived by using a real time clock. However, this method can be imprecise because there might be deviations between the nominal and the actual sample rate, drifts in the two clocks, or buffer overflows/underflows which cause samples to be missed. Furthermore, variable-rate compression schemes, or

---

[4] this could happen anyways even in blocking mode.

[5] we should have really used the name FIONWRITE since this function is very general and not peculiar to audio devices. E.g. it could be useful on tty devices, on network sockets and everywhere we have some amount of buffering in the hardware or the kernel.

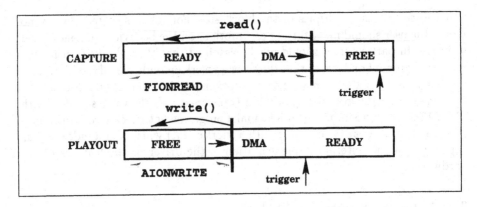

Fig. 3. A view of the capture and playout buffers, with the indication of events which trigger the actions requested with AIOSYNC. The thick vertical bar indicates the current position of the DMA pointer. Also indicated are the effect of read() and write(), and the values returned by FIONREAD and AIONWRITE.

dynamically-changing buffer sizes, might render the use of timers for determining queue occupation completely useless. As an example, it is easy to conceive a device where some functions (e.g. DFT, compression, rate conversion, filtering) are performed in the DSP associated to the CODEC, and the application transfers audio data in a transformed domain or with a variable-rate format. In these cases, deriving timing information is not straightforward.

FIONREAD and AIONWRITE only provide limited information on the status of buffers, and are of use mainly when the application wants to avoid blocking on the device. We would also like to have alternative mechanisms which either notify a process (e.g. using a Unix signal) or wake it up when a specified event occurs. Figure 3 shows in detail the capture and playout buffers. The boundary between the FREE and READY regions, which we call the *current DMA pointer*, moves with time (in opposite directions for the capture and playout channel). Applications can, in general, be interested to know when the current DMA pointer reaches a given position, relative to either the beginning or the end of the buffer, and have the need to be notified when the event occurs, and to know the exact position of the current DMA pointer relative to either end of the buffer.

To support these functionalities we have introduced a single new ioctl(), AIOSYNC, which takes the specification of an event (the current DMA pointer reaching a given position in either buffer) and an action to execute when the event occurs. The event can be specified relative to either end of the buffer, while the action can be any of the following (in all cases, upon return, the current DMA pointer will be reported, relative to the same buffer end as used in the request):

- **no operation.** This function is blocking (unless the event has already occurred) and will return when the desired event occurs. This function is very powerful and flexible; it can be used to wait for a buffer to drain or fill up to a certain level, or just to report the status of the transfer (duplicating to some extent the information supplied by FIONREAD and AIONWRITE, although with AIOSYNC we can read the current size of either the FREE or the READY region of the buffer).

- **send a signal.** This function is non-blocking. It schedules a signal to be sent to the process when the buffer reaches a given mark. This provides an asynchronous notification which can be handled while a process is active, or wakes up a process blocked on a system call.

- **wakeup a selecting process.** This function is non-blocking. It causes a process blocked on a select() call to be woken up if it is waiting for exceptional conditions on the audio file descriptor and the desired event occurs. Note that this action is not an exact duplicate of the previous one: while a signal scheduled with the previous function can wake up a selecting process, there is a potential race condition in that the sequence

```
ioctl(fd, AIOSYNC, ...);
ret = select( ... );
```

might be interrupted in the middle, and the signal be delivered before the select() call. The problem can be solved but at the price of some obfuscation in the code. With this function, we simply request a select() for *exceptional conditions* on the file descriptor (specified using the fourth parameter of select()) to wake up when the desired event occurs. This makes us not affected by timing issues since the event is possibly logged in the device driver and reported to the application as soon as select() is invoked.

AIOSYNC covers all practical needs for synchronization, and the cost of implementing the different notification methods is minimal since they share almost the same code paths.

The resolution of the AIOSYNC calls depends a lot on the features of the underlying hardware. On some devices, the DMA engine can be reprogrammed on the fly to generate an interrupt exactly when the desired event occurs. On other devices, this cannot be done, so if the desired event falls within the boundaries of an already started DMA transfer, there is no alternative but to periodically poll the status of the transfer. In this case, the resolution which can be achieved depends on the granularity of the system's timer, since the poll is generally done once per timer tick. Common values for the timer frequency correspond to a granularity of 1..10 ms, which are acceptable for the coarse synchronization of streams (consider that 10 ms correspond, at the speed of sound, to about 3 meters, comparable to the distance between players in an orchestra; the refresh rate of most video devices is in the 10-20 ms range, so a synchronized video output with higher resolution would be useless; moreover, non real-time operating system would probably make a higher resolution not very useful because of the jitter in scheduling processes).

The last ioctl() we use to support synchronization is AIOSTOP. The function takes the indication of a channel, and immediately suspends the transfer on that channel, flushing the content of the kernel buffer. The return value from the function is the amount of data queued in the buffer when the channel was stopped. This function allows the application to suspend a capture as soon as it decides that no more data are needed, and directly supports the PAUSE function in audio players. It is responsibility of the application to reload any data that was not used in the play buffer. There is no ioctl to start a transfer, since this action is implicit when issuing a read(), write() or select() call.

Function	Description
FIONBIO	Selects blocking or non-blocking mode of operation for the device
FIONREAD	Returns the amount of data which can be read without blocking.
AIONWRITE	Returns the amount of data which can be written without blocking.
AIOSSIZE	Selects character or block mode of operation for the device, setting the threshold for select() to return. A count of 0 or 1 means the usual character mode, a count > 1 makes select return only when the specified amount of data is available. AIOGSIZE returns the size currently in use.
AIOSYNC	Schedules the requested action (return, signal, or enable select) at the occurrence of the specified event. Returns the current status of the buffer.
AIOSTOP	Immediately stops the transfer on the channel, and flushes the buffer. Returns the residual status of the buffer.
read()	Returns at most the amount of data requested. Might return a short count even in blocking mode. In non-blocking mode, it will return immediately with the data already available in the buffer. Also start a paused capture.
write()	Writes at most the amount of data requested. Might return a short count even in blocking mode. In non-blocking mode, it will return immediately after filling up some amount of the free space in the kernel buffer. Also start a paused playout.
select()	In character mode, will return when at least one byte can be exchanged with the device. In block mode, will return when at least a full block (of the size specified with AIOSSIZE) can be exchanged with the device. Also start a paused transfer.

Table 1. Functions supported by our audio driver for data transfer and synchronization.

For reference, Table 1 summarizes all the functions related to synchronization and data transfer supported by our driver. It might surprise that there is no function to set the size of the buffer in the device driver. We do not believe it to be useful, for the following reasons:

- the purpose of the buffer in the device driver is to avoid that applications have to be scheduled too often to communicate with the CODEC. A busy system, or a system where the scheduling clock runs at low frequency, needs larger buffers than one where timeslices are very short. But these parameters are not readily available to applications, so it is the operating system which should decide how big the internal buffers should be;

- being handled by the DMA engine, these buffers must reside in non-pageable memory, and take resources permanently on the system. As a consequence the decision of how much buffer space to use *in the kernel* is not something that a single application can take, but should be taken by the operating system itself depending on the actual resources (total memory, buffers used for other devices, etc.) which are available;

- applications will need to have their own buffering in almost all cases, and they have much greater control on user-space buffers than on kernel-space buffers. In order to write more portable and reliable software, it is much better to force applications to provide their own buffering scheme, suited to the actual needs, than to rely on resources which might not be available to the same degree on all systems, and which impose stronger limitations on their use (e.g. because it is much harder to remove data from the kernel play buffer than from the user-space play buffer).

As a consequence, we left out such a call on purpose. We believe that portability and clarity of programs can improve if they do not depend on being able to set the buffer size in the kernel, and instead provide any required buffering within themselves. These are the same reasons which suggested not to include functions to manipulate the content of the internal buffers of the device driver.

5 Related work

There is unfortunately relatively little published work on audio device drivers. Most work on multimedia devices focuses on video acquisition and rendering, which has more demanding requirements in terms of processing and data-movement overhead. Most operating systems implement a primitive interface to the audio hardware, giving only access to the basic features of the CODEC [1], and with little or no support for synchronization.

The mapping of kernel buffer in the process' memory space has gained some popularity in recent times, on the grounds that this technique can save some unnecessary copies of data [1, 10]. Having the buffer mapped in memory also gives the (false) sensation that programs can gain functionality. As an example, the typical use of mmapped buffers in audio conferencing programs is to pre-initialize the playout buffer with significant data (e.g. white noise, or silence) to minimize the effect of missing audio packets. For games, things can be arranged so that some background music is placed in the buffer and played forever without further intervention.

It is evident that the above are non-problems in a modern system when dealing with audio data. Audio samples have a rate of 192 KB/s at most, whereas the memory bandwidth of modern systems is 2-3 orders of magnitude higher, so the copy overhead is minuscule (for video it would be all a different story). Provided a suitable synchronization mechanism exists, such as AIOSYNC, the pre-initialization of the buffer described above can be easily implemented in the application using the conventional read/write interface, also gaining in programming clarity. Additionally, for special applications such as audio conferencing, pre-filling the buffer can be efficiently done in the driver itself (as we in fact do in our driver). Finally, separate processes or threads can be used to generate background audio even in a more flexible way.

In many systems, access to the audio device is mediated through a library [9] which provides additional functionalities such as mixing multiple streams, playing entire files in the background, etc.. This approach is certainly advisable, although a libraries can only export and simplify the use of functionalities existing in the device driver.

Another popular approach for audio applications is to mediate access to the audio device through a server process [8, 3], similar to the X Window server. The very nature of audio poses however some limitations to this approach. Multiplexing audio output is not as simple as for video, where multiple independent windows can be created. Thus, mechanisms are required to move the "focus" of the server from one application to the other, either manually or automatically. The second, more important, problem is related to the real-time nature of audio: mediating data transfers through an additional process, and possibly through a communication channel, can introduce further, unpredictable, delays in the communication with negative effects on some applications.

6   Discussion and performance

Most of the primitives described in the previous Sections have been implemented in our audio driver [11] for the FreeBSD operating system [2]. The driver supports a variety of audio cards, with different features and limitations. All of the supported cards use the services of the ISA DMA controller to support DMA operations. As a consequence, most of the functionalities described in this paper could be supported by using some simple code to fetch the transfer status from the ISA DMA controller. In order to obtain the asynchronous notifications needed to wake up sleeping processes, two approaches have been followed. In case the audio device supports interrupting a DMA operation on the fly, then the device is reprogrammed to generate an interrupt when the desired event occurs. When this is not possible, a periodic handler is scheduled to process the event within one timer tick from its occurrence. The overhead for the periodic handler is very small, and the resolution is 10 ms with the default timer frequency (100 Hz).

6.1   Applications

We have used our new driver in a number of audio applications. In many cases, source code was available and we could update the applications to make use of the new interface, or simplify the code because the new functionalities were simpler to use.

The AIOSIZE call has been used to improve the behaviour of audioconferencing programs such as vat. vat transfers data in blocks of 160 bytes, corresponding to 20 ms of audio sampled at 8 KHz. The choice of the frame size is related to the compression algorithms (GSM, LPC) used in the program. The main body of the program calls select() on a set of file descriptors, which include the network socket, the X server, and the capture audio device. When operating in character mode, the select() would return after 125 μs, causing the process to consume a huge amount of CPU time just to read one character at a time and loop waiting for a full frame to be available. With block mode operation, we reduced the CPU occupation from about 50% to a mere 3% when not using compression. The advantage here (and in other similar cases) comes mostly from the availability of new synchronization mechanisms.

The AIOSYNC call, has been used in vat to control the length of the READY region of the playout buffer (the same goal can be achieved with AIONWRITE if the buffer size is known). vat, rat and other audio tools try to control the length of the playout buffer basing on the assumption that the capture and playout sections work at exactly the same sample frequency [5, 6, 7]. Under this assumption, the length of the playout buffer can be kept constant if as many bytes are written as they have been read from the capture section. There are situations where the assumption does not hold, e.g:

- when two physically distinct devices are used, sample clocks might be slightly different or drift over time;
- if DMA transfers are disabled while restarting a DMA operation, the average sample rate on each channel depends on how samples are lost/delayed while the DMA engine is restarted.

We have also encountered buggy CODECS which miss samples during regular operations. In all these cases, using the amount of data read as a measure of the speed of the write channel does not help, and in the long term it will cause underruns to occur, or data to accumulate in the output buffer, with consequent clicks or delays in full duplex operation. AIOSYNC provides a way to detect the occurrence of such events, and compensate them. In fact, before issuing a write() operation, the queue length can be read. If the value, in the long term, differs significantly from what is expected, the size of the write is slightly adjusted in the opposite direction to compensate the difference before the error becomes too large.

The asynchronous notification mechanisms provided by AIOSYNC have many other applications. Consider, as an example, a player process being run on a workstation in the background, e.g. doing MPEG decoding in software. During normal activities, the process has plenty of CPU available, and can keep the playout buffer almost full at all times. When the machine becomes loaded (e.g. during a compilation), the buffers might drain with the risk of having pauses in the audio output. The player process could then program a signal to be sent when the buffer reaches some low water mark, and switch to some less expensive operating mode (e.g. mono instead of stereo) upon reception of the signal.

As another application of **AIOSYNC**, a multimedia application might want to launch a free-running animation while some audio description is played on the audio device. The application could then schedule a **signal** to be sent when the buffer becomes almost empty, and start running the animation without worrying about the audio. This is useful since the animation code might use blocking primitives (e.g. while reading data from the disk, or sleeping between subsequent frames), and taking care of the audio would require heavy program restructuring or a separate thread. In our case, instead, the **signal** handler will take care of sending more data to the audio device, or forcing the termination of the animation when the appropriate conditions occur.

Other applications of the **AIOSYNC** calls include games and all those programs where visualizations or other actions (e.g. controlling an external instrument) should be done synchronously with audio events.

7   Conclusions

We have described a software interface for audio devices to improve support for multimedia applications. In defining this software interface we have tried to pursue the following goals:

- look at what could be the real needs of applications, rather than try to extend some existing software interface;
- only specify the external interface of the device driver, do not rely or make assumption on the internal structure of the driver or of the hardware. Do not export information which could lead to non-portable code to be written.
- keep the number of functions small;
- avoid duplication of functionalities in the interface, so that there is no doubt on what is the preferred method to achieve a given result;

We believe we have achieved the above goals, since our interface is small, powerful and simple to use, and resulted in a very compact implementation of device drivers. Our initial experience in porting applications to the new interface has been highly positive, since in many cases software modules which access the audio devices could be largely simplified and in some cases improved by making use of the functions provided by our interface.

Acknowledgements

The work described in this paper has been supported in part by the Commission of European Communities, Esprit Project LTR 20422 – "Moby Dick, The Mobile Digital Companion (MOBYDICK)", and in part by the Ministero dell'Università e della Ricerca Scientifica e Tecnologica of Italy.

References

1. P. Bahl, "The J300 Family of Video and Audio Adapters: Software Architecture", Digital Technical Journal vo.7 n.4, 1995, pp.34-51
2. The FreeBSD operating system Web page, http://www.freebsd.org/
3. J. Fulton, G. Renda, "The Network Audio System", 8th Annual X Technical Conference, in "The X Resource, Issue Nine, January 1994".
4. V.Hardman, M.A.Sasse, M.Handley, A.Watson: "Reliable audio for use over the Internet", INET'95 conference.
5. V.Hardman, I.Kouvelas, M.A.Sasse, A.Watson: "A packet loss Robust Audio Tool for use over the Mbone", Research Note RN/96/8, Dept. of Computer Science, University College London, 1996.
6. V.Jacobson, S.McCanne: "The LBL audio tool vat", Manual page (ftp://ftp.ee.lbl.gov/conferencing/vat/)
7. I.Kouvelas, V.Hardman: "Overcoming Workstation Scheduling Problems in a Real-Time Audio Tool", Proc. of Usenix 1996.
8. T.M. Levergood, A.C. Payne et al., "AudioFile: Network-Transparent System for Distributed Audio Applications", USENIX Summer Conference 1993, June 1993.
9. Microsoft Corp., Documentation on the DirectSound SDK, available at http://www.microsoft.com/DirectX/
10. The Open Sound System (OSS) Web page, http://www.4front-tech.com/
11. L.Rizzo, Sources for the new FreeBSD audio driver, available from http://www.iet.unipi.it/~luigi/FreeBSD.html
12. H.Schulzrinne: "Voice communication across the Internet: A Network Voice Terminal", Technical Report TR 92-50, Dept. of Computer Science, University of Massachusets, Amherst, July 1992.
13. T.Turletti: "The inria videoconferencing system (ivs)", ConneXions – The Interoperability Report, 8(10):20-24, October 1994.

Cell Discarding Mechanisms
with Minimum Flow Quality*

Josep Mangues-Bafalluy and Jordi Domingo-Pascual
e-mail: jmangues@ac.upc.es and jordid@ac.upc.es
Universitat Politècnica de Catalunya. Departament d'Arquitectura de Computadors.
Campus Nord. Mòdul D6. Jordi Girona 1-3. 08034 Barcelona.
Phone: +34 3 4016981   Fax: +34 3 4017055

Abstract. The fragmentation of higher layer packets into ATM cells is a major problem in an ATM network if congestion arises. Once one cell is lost, the remaining cells belonging to the same packet are not useful, and thus, the interest of selectively discarding such cells is the focus of many papers.

When dealing with multimedia real-time interactive services, their stringent QoS requirements demand the use of effective congestion control mechanisms to guarantee a minimum QoS.

The Active Cell Discard (ACD) mechanism and some modifications presented in this paper try to guarantee a minimum QoS by making sure that there is always enough free space in the buffer to queue a set of initial cells of the packet. Even if the tail of the packet is cut out by a cell discarding mechanism, the initial set of cells may be useful to maintain the QoS of the application. It provides a minimum flow of information, and critical information may be delivered in these initial cells of the packet.

In this paper, we present some of the results obtained for the performance of the ACD mechanisms and compare them with other known proposals. The traffic model used for this evaluation follows the characteristics of the ATM traffic measured in the Spanish academic ATM network.

1. Introduction

BISDN based on ATM, are expected to provide different kinds of services with distinct QoS requirements on the same network.

Statistical multiplexing of ATM cells belonging to different connections has the advantage of high resource utilization, but the management of the QoS requirements of a multiplexed flow is very complex. Moreover, the variable behavior of the traffic characteristics can lead to congestion situations if there are not enough network resources to serve all the traffic.

Congestion control mechanisms help in coping with congestion either reacting to these situations or preventing them from happening.

---

* This work has been funded by the Spanish Ministry of Education (CICYT) under grant numbers TIC95-0982-C02-01, TEL96-2509E, and by the Catalonian Research and Universities Council (CIRIT 1995SGR-00464).

When transmitting packet oriented communications over an ATM network, the fragmentation of packets into ATM cells makes congestion more harmful to the connection. Although the Cell Loss Ratio (CLR) is an important parameter, in packet oriented communications the minimum meaningful unit of information for the end-user is the packet. Therefore, the important parameter that has to be minimized is the Packet Loss Ratio (PLR).

Interactive real-time multimedia traffic introduces more stringent conditions to the QoS management functions. In this case, the Cell Transfer Delay (CTD) and Cell Delay Variation (CDV) must be minimized. These requirements have important consequences on the network design. For example, if the buffers at the switching nodes are long, complex CTD and CDV control mechanisms must be used.

The working scenario considers an ATM network where buffering at the switching nodes is dimensioned to cope with cell contention. In this way, as buffers are short, minimal control for CTD and CDV must be implemented while a better congestion control must be used.

The goal of a congestion control mechanism is not only to try to reduce congestion but to guarantee a minimum QoS for the most critical services even in situations of congestion.

In the next section, previous work on cell discarding policies is reviewed. In section 3, the Active Cell Discard mechanism is explained along with some studies related to it. Some variations of the ACD mechanism are introduced in section 4. Next, the simulation environment is explained and in section 6, the results obtained by simulation are shown. Finally, some conclusions are presented.

2. Cell Discarding Policies

Cell discarding policies are one of the congestion control mechanisms used by the network to either avoid congestion or to minimize its effects once it has appeared. Their usefulness relies on the fact that when transmitting packet oriented information over an ATM network, the fragmentation of the packets produces very poor performance in comparison to that obtained over packet switched networks.

In [ROFL 95] two discarding mechanisms (Partial Packet Discard, or PPD, and Early Packet Discard, or EPD) are introduced and some comparisons of performance are presented. In that case, TCP over the ATM layer was used and the influence of its end-to-end congestion control behavior has an important contribution to the results.

Studies with fixed length TCP data packets are presented in [ROFL 95] and in [LAKS 96]. In the former, EPD shows a better performance than PPD. The second one presents a discarding strategy, called Drop From Front, specially prepared to work with TCP versions using fast-retransmit. The performance results show a performance comparable to that of RED (Random Early Detection) with whole frame drop.

In [KAWA 96] some discarding policies are analytically compared without considering any other higher level mechanism. The length of transmitted packets follows a geometric distribution and, for this traffic, Tail Dropping, i. e. PPD, shows a better performance than EPD. This behavior is caused by the variability of the packet length distribution.

Though the geometric distribution is a more realistic approach than fixed length packets, it is still far from the real traffic behavior. That is why it is very important to evaluate these congestion control mechanisms when they are applied to real traffic on ATM networks.

Most of the studies done so far use TCP over the ATM layer. But, there is an important effort to develop new transport protocols to be used in ATM networks. Therefore, it would be much more interesting to implement generic congestion control mechanisms whose performance is not dependent on the protocol being used.

Another remarkable point is that the mechanisms reviewed above were validated with long buffers because they are oriented to deal with data traffic. As a consequence, CTD and CDV must be bounded using additional control mechanisms.

3. Active Cell Discard

The ATM architecture, in its ATM layer offers a bearer service which is time and semantic transparent. Thus, the AAL layer is in charge of adapting the ATM network to the requirements of the services. Bearing this architecture in mind, the aim of this mechanism is to be independent of any congestion control mechanism used by protocols over the ATM layer. Consequently, it must be useful for any kind of service and traffic types. Some cell discarding mechanisms found in the literature are dependent on the higher layer congestion control, for example the Drop From Front strategy [LAKS 96].

3.1 Motivation

The variety of services offered by the BISDN makes necessary the use of congestion control mechanisms as general as possible to accomplish the present requirements and those of future services.

Interactive real-time communications don't allow retransmissions of lost packets. The use of short packets tends to minimize the effect of packet losses due to congestion but introduces significant overheads. Therefore, it is interesting to guarantee that at least a portion of the packet arrives at the receiver and the packet is not completely discarded. The objective is to maintain the connection within the minimum QoS required by the end-user by means of the information carried by this portion of the packet to guarantee a minimum flow. In that way, a minimum cell rate may be guaranteed.

Some reorganization of the information may be needed to put the most relevant information at the beginning of the packet. For example, in the case of video, the set of initial cells could be used to carry the audio information, the synchronism information and some video information to guarantee a minimum image quality. Some kind of scalability will be necessary in the media flow coding process. Furthermore, the AAL should be capable of delivering the set of initial cells as a 'minimal' packet although the remaining of the packet has been discarded.

Moreover, today's protocols present a bi-modal behavior. They usually use long packets for user information and short packets for control purposes. The user information though being important it is not as important as the control one, which is vital to maintain the connection within the desired QoS. The mechanism proposed here would be very appropriate to guarantee that this information reaches the end terminal even at high congestion levels.

Some studies of cell discarding mechanisms without the influence of TCP were presented in [DOM 95a] and [DOM 95b], and a new mechanism, Active Cell Discard (ACD), was introduced in [MAN 97a]. In this latter paper, some comparisons among PPD, EPD, ACD and the case without congestion control were presented.

The environment used to simulate this mechanism was very simple; all the sources transmitted fixed-length packets and had the same characteristics. With that environment, the results of ACD appeared to be very promising, because under severe congestion, the delivery of a minimum set of cells of every packet could be guaranteed with a very low Packet Loss Ratio (PminLR).

The results showed that the PLR of entire packets could be higher than the one obtained with EPD, but in most cases, the PLR of the set of initial cells (PminLR) was some orders of magnitude below [MAN 97a].

A new aspect was introduced in [MAN 97b] where the characteristics of the simulated traffic used to validate the mechanism were chosen according to the traffic measurements obtained in the ATM backbone of the Spanish academic network [CASTBA 97], although it was a rough approximation.

Measurements show a high percentage of short packets (more than 50% fit in less than 3 cells) and medium size packets.

3.2 The ACD mechanism

The contract between the user and the network includes the Peak Cell Rate (PCR), the Cell Delay Variation (CDV) and the number of cells of the packet that must be preserved as the set of initial cells (Pmin).

As in the EPD and PPD, the buffer occupancy (q) must be monitored by the switch, and it should keep the current state per VC. It also needs a counter per VC.

ACD is implemented by using the same principle as AAL5. Multiplexing is not supported in this AAL type and then, the ATM-layer-user-to-ATM-layer-user (AUU) indicator can be used to establish the limits of the AAL-PDU [ATMF 93]. The counter of each VC is not reset until the switch receives the last cell, which should never be dropped.

The main difference between AAL5 and this proposal is that the set of initial cells (Pmin) must be delivered to the upper layer even if the remaining of the packet has been discarded. It may be considered as a modification of AAL5 in which the last cell of the packet (with AUU indication) provides error protection to check the integrity of the set of Pmin cells.

The threshold (Q_t) is used to give enough space for the Pmin cells and to prevent congestion. This threshold is set as a percentage (s) of the queue length, $Q_t=s*Q$ (0<s<1), where Q is the buffer size.

For connections with high Delay*Transmission_rate product, the amount of cells crossing the network is very important. As a consequence, the advantage of preventing congestion rather than acting after it appears is significant. In some way, ACD uses its threshold to prevent congestion from affecting the set of initial cells (Pmin).

The mechanism works as follows:

Be n the packet cell counter value per VC.

1. If $0<n<=Pmin$ and $q<Q$ the cell is not discarded. A Pmin cell is only discarded when there is no space in the buffer. In this case the remaining cells are discarded until the cell with AUU indicating the end of the packet is found.
2. If $n>Pmin$ and $q<=Qt$ the cell is queued.
3. If $n>Pmin$ and $q>Qt$ the cell is discarded, and all the remaining cells of the packet, too.
4. The last cell is queued if there is enough free space in the buffer.

This last point is important because AAL functionality relies on the AUU indication and error detection. For this reason the cell loss probability of the last cell is obtained and shown in the performance results.

As it can be deduced from these four points, the threshold is used to reserve the space between the threshold and the buffer capacity for Pmin cells and last cells only. The selection of the threshold value is a trade-off between the amount of user information lost and the guarantee that none of the Pmin packets is lost. This value depends on the traffic parameters with which the switch will work. More work has to be done in that direction.

4. Active Cell Discard Variations

Some variations of ACD have also been studied. All of them need AUU indication to be implemented. They preserve the initial goal of ACD, i.e. preserving the set of initial cells, and try to improve the PLR for entire packets.

Therefore, the difference between them is the way they treat the cells not belonging to the initial set that must be preserved. The original mechanism, which has been presented in the previous section, uses a partial packet discarding strategy for those cells, we will refer to it as ACD-PPD.

4.1 ACD-EPD

Motivation

Having let the set of initial cells pass, the goal is to manage to get as many complete packets as possible. This second part of the packet (formed by cells not belonging to the set of initial cells) has a significance to the final user if and only if all cells are delivered. Therefore, once one of these cells is lost, we must eliminate the rest of the packet, which is already done by ACD-PPD, but it seems that a better performance would be obtained if we eliminate the second parts as a whole in the same way EPD does with whole packets.

Algorithm

Be n the packet cell counter value per VC.

1. If $0<n<=Pmin$ and $q<Q$ the cell is not discarded. A Pmin cell is only discarded when there is no space in the buffer. In this case the remaining cells are discarded until AUU indicating the end of the packet is found.
2. $n=Pmin+1$ and $q<=Qt$ the cell is queued otherwise it is discarded and the remaining cells of the packet, too.
3. If $Pmin+1<n<P$ and $q<Q$ the cell is queued.
4. If $n=P$ (last cell), it is queued if there is space in the buffer and at least one cell of the present packet has already been queued.

4.2 ACD-EPD2

Motivation

Only using a threshold, as in ACD-EPD, could result in an increased PLR of the initial cells because once the second part of packet is accepted, the cells belonging to it have the same priority as the initial cells, i.e. the excess buffer capacity (from Qt to Q) is not only filled by Pmin cells as in ACD-PPD but also by not initial cells whose second part was accepted.

As a consequence, another threshold (Qe) is introduced with the idea that even if the second part is accepted, there will always be a part of the buffer just used by Pmin cells. And the region between both thresholds will be used by cells belonging to second parts of the packet. EPD is applied when the buffer occupancy is greater than Qe.

Algorithm

Be n the packet cell counter value per VC.

1. If $0<n<=Pmin$ and $q<Q$ the cell is not discarded. A Pmin cell is only discarded when there is no space in the buffer. In this case the remaining cells are discarded until AUU indicating the end of the packet is found.
2. $n=Pmin+1$ and $q<=Qe$ the cell is queued, otherwise it is discarded and the remaining cells of the packet, too.
3. If $Pmin+1<n<P$ and $q<=Qt$ the cell is queued, otherwise it is discarded and the remaining cells of the packet, too.
4. If $n=P$ (last cell), it is queued if there is space in the buffer and at least one cell of the present packet has already been queued.

The buffer positions ranging from Qe to Qt are used by second part cells, while the positions from Qt until Q are occupied by Pmin cells.

4.3 ACD-RED

Motivation

Another interesting approach comes from what has been observed in [LAKS 96] where Random Early Detection (RED) with whole packet drop showed a very good performance.

Like in [LAKS 96] we chose an exponential distribution as drop function, but with the consideration that there must always be a region in the buffer only used by Pmin cells. Then, the drop function presents an exponential distribution until the threshold (Qt) and its value is 1 for the remaining buffer positions. In this case no threshold management is needed because it is implicit in the drop function.

Therefore, in this variation, the RED algorithm is applied for the second part of the packet (cells following Pmin cells).

Algorithm

Be n the packet cell counter value per VC.
1. If 0<n<=Pmin and q<Q the cell is not discarded. A Pmin cell is only discarded when there is no space in the buffer. In this case the remaining cells are discarded until AUU indicating the end of the packet is found.
2. n=Pmin+1, it is discarded with a probability indicated by the drop function. Once this cell is discarded the remaining cells of the packet are not queued.
3. If Pmin+1<n<P and q<Q the cell is queued.
4. If n=P (last cell), it is queued if there is space in the buffer and at least one cell of the present packet has already been queued.

5. Simulation Environment

All the mechanisms are simulated in a single switch system with a very low buffer capacity (Q=20 cells). The reason for such a small buffer comes from the traffic characteristics of the interactive real-time multimedia flows this mechanism is expected to deal with. The aim is to reduce the CDV and the CTD appearing in a system with a lot of hops.

The buffer is shared by many inputs and because of its length, it is only able to cope with cell level congestion but not the burst level congestion [DOM 95b]. We also consider that the physical links are perfectly reliable and the packet losses are just due to buffer overflow.

Neither higher level congestion control nor protocol are considered. Therefore the PLR results are only due to the behavior of the simulated mechanism. Thus, we obtain general conclusions, that are independent from the protocols over the ATM layer.

The simulated mechanisms are those described in the preceding section and, in general, the requirements of the switch for all of them are:

- per-VC signaling
- inspection of the Payload Type (PT) to determine if it is the last cell in the present packet
- buffer occupancy monitoring. Depending on the mechanism there are two options :
    1. If a threshold is used, the buffer occupancy can be over or under Qt.
    2. If the mechanism uses a drop function, like in RED, the discarding probabilities have to be associated to each buffer position.
- storage of Pmin, the length of the set of initial cells to preserve. This may be specified on a per-VC basis.
- counter of the packet cells, whose value determines the treatment received by the cell. Reception of AUU cell resets the counter.

Although each source specifies a Pmin value in its contract that may be different from that of the other sources, in our simulations, this value is the same for all the sources.

The number of input sources to the switch is calculated as a function of their traffic characteristics so that the desired global load is achieved.

5.1 Traffic Model

The reviewed studies were based either on fixed packet lengths or on packet length distributions chosen to be treated analytically. These assumptions could lead to partial results because of not being close to a realistic traffic. This is the reason why we use the measurements made on a real ATM network.

Due to the lack of information providing a characterization of multimedia traffic over native ATM networks, we decided to evaluate the ACD mechanism with the traffic that can be presently found in a real ATM network. It is IP traffic which transports all kinds of services, including IP multicast. MBONE traffic generated in videoconference or audio transmission could be considered as multimedia traffic. But, at the same time, due to the bi-modal behavior of the classical protocols (such as telnet or ftp), ACD proves to be useful for this type of information, too.

5.2 Real Traffic Measurements.

In project CASTBA [CASTBA 97], some measurements of the ATM traffic traveling through the Spanish academic ATM network were performed. This network uses IP over ATM.

Some of the measurements are the distribution of packet lengths, the mean overhead traffic, and the distribution of the traffic load during a certain period of time.

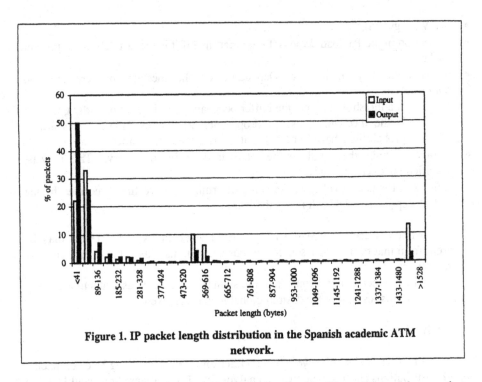

Figure 1. IP packet length distribution in the Spanish academic ATM network.

If we look at the mean packet length distribution (figure 1), we observe that most of the traffic can be mapped into CS-PDUs of three or less ATM cells. Two different distributions are observed for input traffic (traffic going from the network backbone towards a local node) and for output traffic (in the other direction). This different behavior appears because of the asymmetrical nature of the network. Local nodes act as information consumers. Information is transmitted towards the node in long packets. But, the output direction is mainly used to control the information transfer and to ask servers for the information requested by means of short packets.

Roughly, this distribution can be divided in three groups:
1. Short packets, which mainly correspond to TCP control packets or to telnet packets.
2. Mid-sized packets correspond to the mechanisms used by TCP to minimize the segmentation. In some TCP implementations the MSS (max. TCP segment size) must be a multiple of either 512 bytes or 536 (TCP default), which gives the two peaks in the middle (when the IP and TCP headers are added). Also around these values, there is the minimum recommended IP MTU (576 bytes).
3. Long packets correspond to the 1500 bytes of the Ethernet MTU.

If we consider, for example, the measurements of an interactive service such as telnet, almost the 90% of the traffic can be mapped into 2 cells or less, because of the short packet lengths. Other protocols, such as ftp or http, show a distribution similar to the mean, represented in figure 1. IP multicast presents a slightly different packet length distribution. However, most of the packets (about 68%) can be mapped into 5 cells or less.

However, there are other protocols showing a distribution being very far from the mean, e.g. nntp which uses a lot of long packets and very few short packets in the input direction.

Another important measurement is the percentage of overhead introduced by TCP, IP and ATM, whose mean is about 18.21% of the total traffic. This overhead is specially important for short packets which have to be transmitted in few ATM cells. For example, if an IP packet with 41 bytes is transmitted, two ATM cells are needed and the overhead introduced is 159%. Therefore, even if we are not considering real-time traffic with very stringent requirements, if this kind of cells is guaranteed for data traffic, the wasting of network resources could be reduced.

5.3 Simulated Traffic Model

According to the packet length distribution of the measurements, we propose the following traffic model:

Each source is modeled by an ON-OFF model, which in the ON state generates packets with lengths according to the following distribution.

Cells per packet	Probability
1	.25
2	.35
12	.25
32	.15

This distribution has a mean of 8.75 cells per packet.

The parameters that characterize each source are:

- P: The packet length, i.e. the number of ATM cells per packet. It follows the above distribution.
- Pmin: The number of cells of the initial set which have to be preserved.
- K: The transmission rate of the source, which corresponds to the inverse of the PCR normalized to the link capacity.
- a: The load offered by the source referred to its maximum capacity (a=1) that would be obtained if we transmitted one cell every K cell slots during all the connection. This latter case corresponds to a CBR source. Then, this parameter measures the burstiness of the source.

For instance, considering a link capacity of 150 Mbps and a CBR source at 10 Mbps, the corresponding values for the parameters are K=15 and a=1. And a VBR source with a PCR of 10 Mbps and a Mean Cell Rate of 5 Mbps would have a=0.5.

6. Results

In this section we present the results obtained for different source and system parameters for the mechanisms described in sections 3 and 4.

The global load of the system is very high (ranging from 88% to 100% of the link capacity). Therefore, we test the system under very severe congestion situation.

As a general result, we observed that the CLR may be very high, but the important parameters (PLR and PminLR) are always below the PLR of EPD, which in other studies showed the best performance.

All the results presented here were obtained for a buffer capacity of 20 cells. Some of them were repeated for longer buffers but the general behavior was the same with an improvement of the PLRs because of the higher buffer length.

6.1 ACD-PPD

In figure 2, we present the comparison of ACD-PPD (which has been described in section 3 as the generic ACD mechanism) with EPD for Pmin=11, K=10, a=.5, and Qt=0.5 (i.e. the threshold is in the middle of the buffer).

The results show that EPD, which in [ROFL 95] presented the best performance has a worse PLR than the one obtained with ACD-PPD for entire packets. And the PminLR (the PLR for the set of initial cells) is well below both of them. This different behavior compared to other studies may be explained by considering the traffic characteristics. In fact, with a Pmin=11, we intend to preserve three of the four types of packets we are transmitting, which represent the 85% of the total traffic entering the switch. The only packet having the most part of it out of the set of the initial cells is the 32 cell packet. Moreover, for ACD-PPD the threshold is in the middle of the buffer, preserving half of it to queue just cells belonging to the initial set.

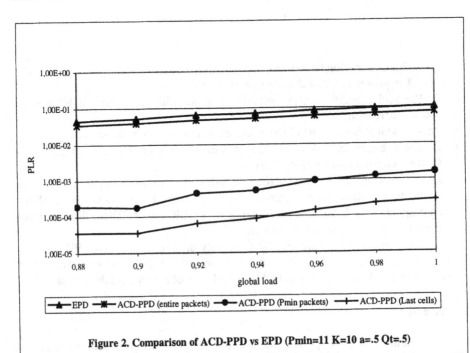

Figure 2. Comparison of ACD-PPD vs EPD (Pmin=11 K=10 a=.5 Qt=.5)

The goal of the ACD-PPD mechanism is accomplished, because the PminLR is two orders of magnitude below the PLR for entire packets with severe congestion in the network.

Last cells are very important for the correct performance of the mechanism, that is why they are treated like the Pmin cells. The results show that the last-cell losses are usually one order of magnitude below the PminLR in case of strong congestion.

Other simulations have been carried out varying K and Qt. All of them showed the same performance and the mechanisms reacted as it could be expected.

When K is higher, meaning that the transmission rate of the sources is lower, the network is loaded with less aggressive traffic and the behavior of all of the mechanisms is better. In the case of ACD-PPD, the difference between PLR and PminLR increases.

When Qt is higher, there is more space in the buffer for cells not being a Pmin cell and the PLR is slightly improved. But the difference is not very important because the 32 cell packets only represent the 15% of the traffic. Increasing the threshold also implies reducing the reserved space in the buffer for Pmin cells, therefore the PminLR becomes worse.

Different Pmin values were also introduced. The results obtained with K=10, a=.5, and Qt=.8 for severe congestion are presented in figure 3.

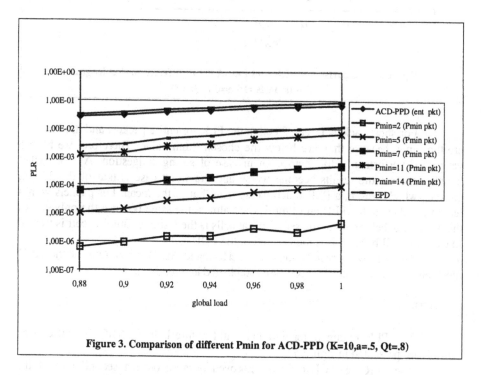

Figure 3. Comparison of different Pmin for ACD-PPD (K=10,a=.5, Qt=.8)

As observed in the figure, the PLR (entire packets) for EPD is always higher than that obtained for ACD-PPD. The PLRs obtained for ACD-PPD in all cases present almost no variation, for this reason, only one line is represented. But the

PminLR is improved considerably as Pmin is reduced because there are less cells to be preserved for the same reserved buffer space.

The behavior of the mechanism as a function of the packet length for Pmin=11, a=0.7 and Qt=0.8 is presented in figure 4. The mean PLR for all packet lengths is also represented for comparison purposes.

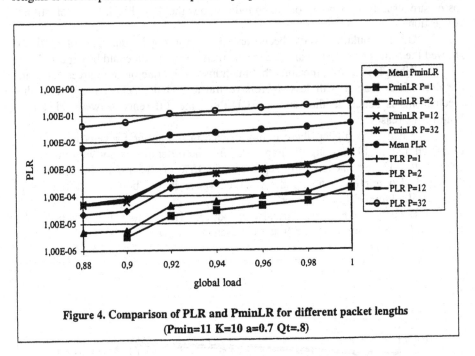

**Figure 4. Comparison of PLR and PminLR for different packet lengths
(Pmin=11 K=10 a=0.7 Qt=.8)**

For short packets (1 and 2 cells), the PLR for entire packets and the PminLR are the same, because they are below the chosen Pmin value. Therefore, we have a very good performance for them even in case of strong congestion. As the packet length increases the results become worse and the worst case is that of the 32 cell packets, which present a PLR much worse than that of the mean. For packets with a length of 12 cells, the PLR and PminLR are almost the same because the only cell of the packet not belonging to the set of initial cells is the last one. But this cell is treated like a Pmin cell because it is the last cell of the PDU.

The following results show the performance of the variations of the ACD mechanism and some comparisons are established among them.

6.2 ACD-EPD

The PLR obtained for Pmin=11 and a threshold of half the buffer size (Qt=.5) are presented in figure 5.

The PLR for ACD-EPD is improved because once a second part of the packet is accepted, the cells belonging to it have the same treatment as the Pmin cells and therefore they are better preserved. But this has a negative effect on the Pmin cells which in the case of ACD-EPD have a PminLR one order of magnitude higher than that of ACD-PPD for these source and buffer parameters. Other results for different parameters show a similar behavior.

If the threshold is higher, the PminLR becomes worse, for almost no improvement of PLR for entire packets.

In both ACD cases being compared the last cells must also be preserved. The losses obtained are almost one order of magnitude below PminLR of its respective mechanism.

We may conclude that ACD-EPD provides no real improvement because PminLR is worse than using ACD-PPD.

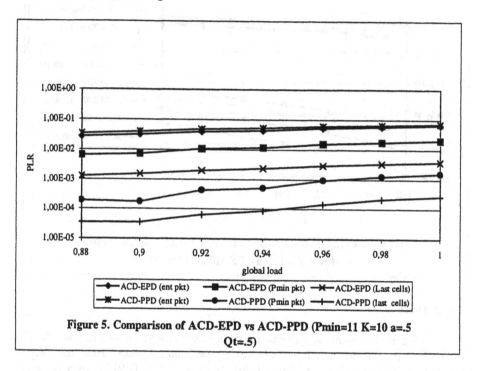

Figure 5. Comparison of ACD-EPD vs ACD-PPD (Pmin=11 K=10 a=.5 Qt=.5)

6.3 ACD-EPD2

Figure 6 shows the performance results for Pmin=7 varying Qe and with Qt=.8 for ACD-EPD2 and the same parameters with a threshold set to the 80% of the buffer size for ACD-PPD.

When Qe is decreased maintaining the same Qt, the PminLR improves considerably. In this case, the best results for PminLR were obtained for ACD-EPD2 with Qe=.1 and Qt=.8 (ACD-EPD2 .1 (Pmin pkt) in figure 6). This is so because with a lower Qe more second parts are eliminated and there is more space in the buffer for the Pmin cells. But the improvement in PminLR implies a worse PLR, having its worst value for Qe=.1 and Qt=.8 in this example, which is the same case that gave the best PminLR.

None of the threshold combinations for ACD-EPD2 attained a PLR as low as that of ACD-PPD, though they are very similar. Therefore, the initial objective of the mechanism was not accomplished, because we just obtained an improvement of PminLR and not of PLR as it was expected.

If the global load of the system is lower, the same behavior is obtained.

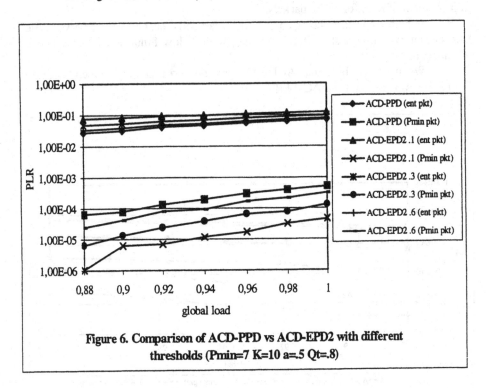

Figure 6. Comparison of ACD-PPD vs ACD-EPD2 with different thresholds (Pmin=7 K=10 a=.5 Qt=.8)

6.4 ACD-RED

Figure 7 presents a performance comparison among ACD-RED and ACD-PPD. The parameters are Pmin=11, K=10, a=.5 and Qt=0.8. And the factor controlling the form of the exponential drop function for ACD-RED has a value of 1.5. This drop function has an exponential form from 1 (1 cell in the buffer) till Qt (where its value is 1).

For the rest of the positions (from Qt to Q) the drop function value is 1.

The results obtained show that this mechanism doesn't improve the behavior of ACD-PPD. There is not any improvement in the PLR in comparison to ACD-PPD and the PminLR is worse. This behavior may be due to having a too short buffer where the effect of the drop function doesn't have a significant contribution to the performance of the mechanism. This is confirmed by the results obtained when varying the exponential form of the drop function, where the variation between the PLR curves was not significant.

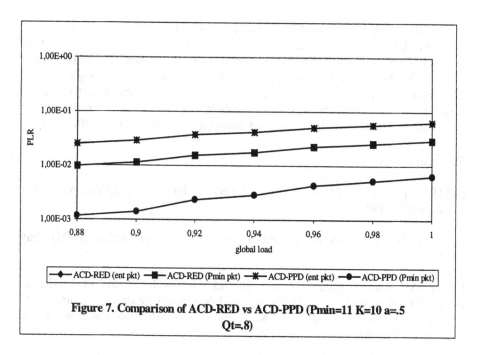

Figure 7. Comparison of ACD-RED vs ACD-PPD (Pmin=11 K=10 a=.5 Qt=.8)

The results were also compared to those of RED with whole frame drop [LAKS 96]. This latter mechanism improves the PLR in comparison to EPD but it is still worse than the PLR for ACD-PPD and ACD-RED for this kind of traffic and for severe congestion.

7. Conclusions

We have shown that the ACD mechanism performs very well in case of severe congestion and for the traffic profile corresponding to IP over ATM (measured in the Spanish academic ATM network). The packet loss probability for the set of initial cells may be some orders of magnitude below the PLR for entire packets.

This congestion control mechanism doesn't depend on the protocols over the ATM layer (i.e. transport protocol), where it is implemented.

The ACD mechanisms guarantee a PLR for the set of initial cells (PminLR) which is much lower than that of entire packets for the methods found in the reviewed literature. In that way, a minimum flow of cells may be guaranteed. And it may provide a minimum QoS to the end-user even under strong congestion and for short buffers which may make it useful for multimedia interactive real-time traffic. Data traffic, such as IP over ATM, may also benefit from this mechanism because it may improve the overall PLR.

The results show that ACD-PPD performs the best if we consider the behavior of PLR and PminLR and the complexity in the implementation.

The comparisons of ACD-PPD and ACD-EPD showed in some cases a little improvement in PLR for entire packets in ACD-EPD, but the PminLR was much worse than that of ACD-PPD.

The comparisons of ACD-PPD and ACD-EPD2 showed an improvement of the PminLR in ACD-EPD2, but the PLR for entire packets was always worse than that of ACD-PPD. The ACD-RED mechanism was not better than ACD-PPD either.

The ACD-PPD mechanism seems to be very promising and simple. Further research with other measured real traffic must be done to be able to determine the optimum values for the Pmin and the threshold (Qt) and to compare it with the performance obtained by other congestion control mechanisms for that traffic.

8. References

[ATMF 93] The ATM Forum, ATM User-Network Interface Specification, Version 3.0, September 1993.

[CASTBA 97] CASTBA Report. Polytechnic University of Madrid (UPM). July 1997.

[DOM 95a] Domingo-Pascual J. and Sánchez-Valdivia J. 'Quality of Service of Packet Data Streams over ATM Networks'. International Teletraffic Seminar. ITC Sponsored. Editors: Goldstein B., Koucheryavy A., Scheneps-Schneppe M. St. Petersbourg (Russia): 404-414, July 1995.

[DOM 95b] Domingo-Pascual J. and Sánchez-Valdivia J. 'Selective Cell Discarding Mechanism for a Packet Oriented Service on ATM Networks'. Computer Architecture Department. Universitat Politècnica de Catalunya (UPC). Research Report UPC-DAC-1995-47. December 1995.

[KAWA 96] Kawahara, K., Kitajima, K., Takine, T., and Oie, Y. 'Performance Evaluation of Selective Cell Discard Schemes in ATM Networks'. Proceedings of the IEEE INFOCOM, Vol. 3: 1054-1061, March 1996.

[LAKS 96] Lakshman, T.V., Neidhardt, A., and Ott, T. J. 'The Drop from Front Strategy in TCP and in TCP over ATM'. Proceedings of the IEEE INFOCOM, Vol. 3: 1242-1250, March 1996.

[MAN 97a] Mangues-Bafalluy J. and Domingo-Pascual J. 'Active Packet Discard Mechanism'. Proceedings of the IEEE Yuforic'97 pp. 119-127, Barcelona, April 1997.

[MAN 97b] Mangues-Bafalluy J. and Domingo-Pascual J. 'Active Cell Discard Mechanism in ATM Networks'. To be published in the Proceedings of DCCN'97, Tel-Aviv, November 1997.

[ROFL 95] Romanow A. and Floyd S. 'Dynamics of TCP Traffic over ATM Traffic'. IEEE Journal on Selected Areas in Communications 13 (4): 633-641, May 1995.

Implementing the Digital Storage Media - Command and Control (DSM-CC) Standard in the Framework of an On-Line Service

Wolfgang Ruppel

Deutsche Telekom Berkom GmbH
Postfach 10 00 03
D-64276 Darmstadt, Germany

E-Mail ruppel@berkom.de

Abstract: Interactive Video Services require interoperable and open server structures. In this paper, a migration path is shown which augments an existing on-line service with broadband Services-on-Demand. The server structure is DAVIC-compliant in order to allow for the integration of different end-user equipment (Internet clients and DAVIC Set-Top-Boxes). The server-side implementation of DSM-CC is described and compared to a proprietary solution.

1 Introduction

Individual interactive broadband services require real-time transport of broadband signals to the subscriber-side terminal. At present, two major methods of video signal transmission are employed: Low-quality video transmission (H.263, MPEG-1/-2<2

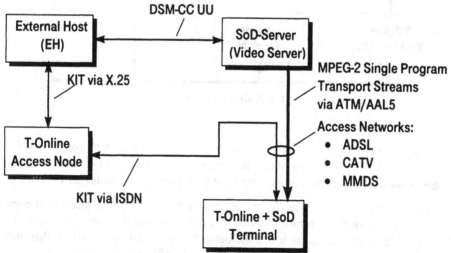

Fig. 1 Demonstrator of the IVES project (block diagram)

Mbit/s) via UDP/IP and „raw ATM" for high-quality video transmission (up to approx. 6 Mbit/s in the consumer domain, MPEG-2 4:2:2 profiles for professional applications). In the latter case, two different ATM adaptation layers (AAL) are used.

AAL5 is supported by the computer industry, especially US companies. AAL1 is preferred for long-distance traffic and server networking because it possesses error correction and detection capabilities. A detailed comparison of the AAL1 and AAL5 layers can be found in [1].

An open Services-on-Demand (SoD) system based on T-Online, Germany's most popular on-line service, has been studied under the IVES project launched by Deutsche Telekom [2,3]. Its basic idea is to complement the existing narrowband service „T-Online" by a broadband offer. In the process, authentication, billing and navigation take advantage of the well-tried T-Online technology. The broadband services spectrum of the T-Online-operated SoD system can be accessed via a so-called „external host" (EH, **Figure 1**). In the EH the protocols used in T-Online are mapped onto protocols conforming to international SoD standards (DSM-CC). The access networks covered by the project are the standard twisted pair telephone wire with ADSL techniques, the CATV network with a dedicated back-channel implemention and a wireless solution using MMDS [4].

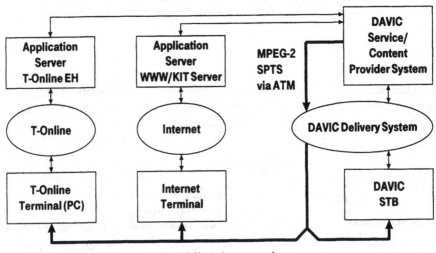

Fig. 2 Migration scenarios

The server network chosen for the IVES project has to be DAVIC-compatible (see Section 3.1). This will allow future migration to Web-based or DAVIC-compatible subscriber terminals without the necessity of new server investments (see **Figure 2**).

Current video servers frequently lean on proprietary software interfaces and partly proprietary downstream formats, so that they are unsuitable for running in a DAVIC-compatible environment. For use in the demonstrator of the IVES project, therefore, DAVIC-compatible interfaces based on proprietary APIs were designed.

2 API structures of existing video servers

Table 1 shows a classification of video servers according to their software interfaces. Proceeding from a closed system (category 1), the software interfaces of which are disclosed at no point, another three categories have been defined.

The term „middleware" as used in this context designates the communication layer(s) between an application and the network. Well-known middleware implementations are, among others, CORBA (Common Object Request Broker Architecture, see Section 3.3), SUN RPC (Remote Procedure Call) and RMI (Remote Method Invocation), the latter forming an integral part of Java. The term „Messaging Protocol" stands for the entirety of control commands exchanged between client and server of an SoD system. The DAVIC-adopted messaging protocol for SoD systems comprises DSM-CC User-to-Network and User-to-User Commands (see Section 3.2). Commercial systems for Video-on-Demand applications fall exclusively into categories 1 to 3. The author knows no complete server-side CORBA-based DSM-CC implementations for VoD applications.

Category	Designation	Description
1	Closed system	Only dedicated client platforms are supported, and exclusively high-level authoring tools available. Messaging protocol and middleware not disclosed.
2	Closed system, proprietary but disclosed messaging protocol	Proprietary, non-disclosed middleware/disclosed messaging protocol: Client APIs available in binary form. Advantage: If API functionality is sufficient, clients can be written themselves.
3	Closed system, open middleware	Middleware conforms to open standard/messaging protocol disclosed but proprietary: Integration of different client platforms possible if the middleware to be used (e.g., ONC RPC or CORBA) is available for the respective platform.
4	Open system	Middleware and messaging protocol conform to open standards: Depending on the middleware used, any client platforms are integrable. Compatibility between terminals from different suppliers, therefore independence of proprietary solutions.

Table 1 Classification of video servers according to software interfaces

The reason for this is, for example, that the existing systems were developed at a time when the DAVIC 1.0 Specification had not yet been adopted. In the light of the still subdued euphoria about VoD services, a wide variety of DAVIC-compatible server products are therefore not to be expected in the short run.

3 Standards for Video Server Software Interfaces

3.1 DAVIC Specification

The Digital Audio-Visual Council (DAVIC) pursues the goal of specifying systems for interactive video services. DAVIC Specifications 1.x [5-8] are invariably compatible with their forerunner versions. The focal points of the individual DAVIC phases are as follows (selection):

- DAVIC 1.0: Definition of a system reference model consisting of interfaces, information streams S1 to S5 and protocols. A distinction is drawn between „High-layer", „Mid-layer" and „Lower-layer" Protocols. The system components are divided into the subsystems „Content Provider System", „Service Provider System", „Delivery System" and „Service Consumer System".

- DAVIC 1.1 (September 1996): Multicasting, Server MIB, MMDS, ADSL, „Cable Modem", Distributed Servers, Java Virtual Machine.

- DAVIC 1.2 (December 1996): Internet Access, Copyright Protection, Watermarking, 3-D graphics, High-quality Audio and Video.

- DAVIC 1.3: Home Networks, Multicast technologies, System Management.

3.2 Digital Storage Media Control Commands (DSM-CC)

The DSM-CC standard [10] defines protocols and interfaces enabling control of MPEG-1 and MPEG-2 bit streams.

In the DSM-CC model, bit streams flow from servers to clients. Both servers and clients are called „users". Signalling between client and network and between client and server is termed „User-to-Network" (DSM-CC UN) and „User-to-User" (DSM-CC UU) signalling, respectively.

In general, DSM-CC UN is useful for the exchange of control information, whereas DSM-CC UU is intended to the exchange of session-related information. Both are independent from each other but DAVIC specifies both to be implemented in a compliant system [5].

The DSM-CC UU interfaces are defined in IDL, i.e. Interface Definition Language (see next section).

A more comprehensive introduction into DSM-CC can be found in [9].

3.3 CORBA as Middleware

DSM-CC specifies CORBA as a possible middleware for a DSM-CC UU implementation. DAVIC prescribes a CORBA-based implementation. For a DAVIC-compatible video server, the DSM-CC software interfaces have therefore to be realized in conformity with CORBA.

A detailed description of the fundamentals of CORBA would go beyond the scope of this paper. This is why only some implementation-relevant aspects will be discussed below.

In a CORBA system, *interfaces* rather than implementations are specified. As in all object-oriented approaches, the basic idea is to provide the functionality of objects via interfaces. CORBA interfaces are described in IDL, that is, *Interface Definition Language*. If a client intends to use a CORBA server, it needs to know its IDL interface only. Then a so-called „object reference" is generated in the client - that is, a reference to a server object generally implemented on a remote computer.

In the case of DSM-CC, for example, a `Stream` Object is defined. The interface of a `Stream` Object consists of methods (functions) allowing a video recorder-like interaction.

The interface known to the client completely encapsulates the implementation proper of a CORBA object. Implementation takes place in a high-level language, where so-called „mappings" of IDL are specified on C, C++, Java, etc. According to the encapsulation principle, client and server, of course need neither to be implemented in the same high-level language nor under the same operating system.

Another essential part of the CORBA Specification are the so-called „CORBA Services". Only in combination with these CORBA services are non-trivial distributed applications using directory services, security services, etc. possible.

The implementation of DSM-CC requires, for instance, the „CORBA Naming Service" (see Section 4.2). CORBA is not the product of a *single* manufacturer but a standard implemented more or less completely by different suppliers. For the project described in the following, a Public-Domain ORB available in source code was used, this being the only ORB we know that exhibits a Naming Service on all participating operating system platforms.

4 Implementation of DSM-CC User-to-User Commands

4.1 Constraints

The aim is to integrate two commercially available video servers into the IVES SoD system. One server product is classifiable into category 2, the other into category 3 (see Section 2). This paper describes the integration of the server according to category 3. The hardware platform is a one-processor Digital Equipment AlphaServer 4000 running Digital Unix 4.0. The machine has a disk capacity of about 40 GB and is able to output 12 streams with a bitrate of 6 Mbit/s each. The underlying software platform for our DSM-CC UU implementation is Digital Mediaplex Version 2.3c.

4.2 Implemented DSM-CC interfaces

On the server side, the DSM-CC User-to-User Interfaces were implemented first. These comprise the ServiceGateway, Directory, Session and Stream Interfaces defined in [10]. Within our project, there is no need for a File Interface since application download is done via T-Online. Therefore, the File Interface has not been implemented yet.

Protocol conversion between the KIT (Kernel Software for Intelligent Communication Terminals) Standard used in T-Online [11] and the DSM-CC takes place in the external host. KIT is a subset of the VEMMI Standard [12]. Every KIT application has a so-called KIT Server Process of its own assigned to it in the EH (**Figure 3**). The KIT Server Process is capable of calling up „application-dependent processing functions" to control the video server via RPCs. These are designed as RPC Servers. The KIT Commands of the terminal equipment, used for control of the video server, are converted by the RPC Server to method calls of the corresponding object references. During each session, a UNIX process acts as CORBA client, which administrates the object references received. The retrievals of the KIT RPC Server

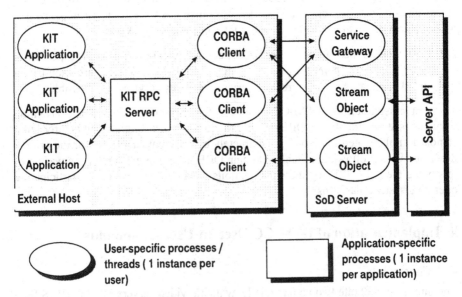

Fig. 3 Process Structure of DSM-CC Communication

are processed separately to avoid bottlenecks. A typical VoD session runs as follows:

After the KIT VoD application has been started, the ServiceGateway:: attach() method is invoked. To pass on client-specific authentication data, use is made of Argument savedContext. This contradicts the semantics of ServiceGateway attach() but is the only way of delivering user-specific data. A movie selection of the customer is converted to a ServiceGateway open() invocation.

In reply, the client receives a reference to a `Stream` object in the server. This `Stream` object implements the functions needed for control of the video (`pause()`, `play()`, etc.). A `Stream` object is closed by means of `Stream close()`. Afterwards, a new stream can be opened with `ServiceGateway::open()` or the session be terminated by entry of `ServiceGateway::detach()`.

The `Stream` service is implemented on top of the server's internal API's which have been made available under a non-disclosure agreement.

In order to introduce platform abstraction the `Stream` service implementation creates an instance of a class `IvesMovie`. The methods of this class implement the methods corresponding to the methods of the `Stream` interface.

When a Stream Service has to be finished by the server, e.g. at the end of a movie, RPC communication takes place between the video server and KIT application process, the syntax of which leans on DSM-CC User-to-Network Commands.

4.3 Selection of the ORB

The Object Request Broker (ORB) must fulfill the following constraints:

- CORBA 2.0 compatibility,

- Naming Service availability,

- IIOP (Internet Interoperability Protocol) implementation.

On evaluation of the ORBs available at present for the target platform (Digital Unix 4.0), the choice fell on a Public-Domain ORB (omniORB) [13] being the only one to meet all the above-mentioned requirements. In a restrictive sense, the current version of omniORB neither supports as yet Data Type Any nor Typecodes, which required a modification of an argument of the `Directory::get()` and `Directory::put()` methods. In addition, omniORB currently supports only persistent servers - a fact that did not impose any restrictions.

Owing to its consistent Multi-Threading Architecture, the omniORB product stands out by high performance and exhibits appreciably shorter response times than commercial ORBs [14].

4.4 Outlook: Distributed server system

After completion of the DSM-CC implementation on the second video server in the IVES SoD system, both servers are intended to form a *single* logic SoD server, which is easily supported by DSM-CC. A further point of study is the implementation of DSM-CC User-to-Network Commands serving session control purposes.

4.5 Results

As became apparent, the DSM-CC User-to-User Commands do not provide a mechanism for call set-up. According to the DSM-CC philosophy, this task is to be performed by the User-to-Network Commands. At the present construction stage, a DSM-CC UN-based proprietary mechanism is used in place of DSM-CC UN. Therefore, the semantics of a parameter (see Section 4.2) had to be modified slightly to allow transmission of client-specific data.

Apart from this aspect, it was found that DSM-CC UU provides highly efficient and well-devised software methods for video server control. A possible obstacle to be overcome may be the availability of a fully functional ORB on the potential target platforms.

5 Comparison with a Proprietary Messaging Protocol

5.1 Implementation based on ONC RPC with Proprietary Messaging

Prior to the development described in Section 4, communication between the EH and video server was implemented in a first step on the basis of a proprietary messaging protocol. The advantage of this solution consisted, above all, in its short-term viability, because a non-MPEG-2-based video server using this protocol had been integrated into the system.

5.2 Merits of an Open Solution

In addition to the general cons in terms of migration toward DAVIC-compatible terminals, an implementation based on proprietary signalling possessed the major disadvantage that server networking was excluded on structural grounds. As a result, this approach was restricted to application in a demonstrator with a small number of subscribers.

However, the solution based on DSM-CC offers all possibilities of networking and integration in a distributed server system. In this case, the physical network structure is invisible and of no importance to the client because only references to objects in the network, identified unambiguously by an object reference in the form of an ASCII character sequence, are passed on.

5.3 Performance Issues

Performance comparison between the proprietary and the DAVIC-compliant messaging has been carried out qualitatively only. There is no remarkable difference for the end-user between the two schemes. Due to its scalability the DAVIC-based solution is the more future-oriented one. It provides inherently the means for adding server resources transparently to the system and managing streaming resources efficiently.

6 Summary

According to the experience gained so far with proprietary solutions, the design of future-safe systems used for interactive broadband services requires due consideration and implementation of international standards. DSM-CC, which has been adopted by DAVIC, provides such an open standard for describing interfaces of SoD servers. Following the basic principles of object orientation, DSM-CC defines only interfaces rather than their implementation. DSM-CC can therefore be realized on different hardware platforms, independently of the internal structure of a video server.

If implementation is not undertaken by the server manufacturer, as in the case described here, the disclosure of software interfaces used for server control is absolutely necessary.

The DSM-CC implementation discussed previously has shown that it can be carried out on a concrete platform. The server obtained in this manner is usable through a protocol conversion in a T-Online-based SoD system of Deutsche Telekom, on one hand, and in future SoD systems with DAVIC-compatible terminals, on the other hand.

7 References

[1] Uhlig, D.: „Übertragung audiovisueller Signale in Multimedia-Kommunikationsnetzen", Proceedings of the „7. Dortmunder Fernsehseminar", ITG 1997. (in German)

[2] Ruppel, W., Breide, S., Kroeger, B.: „Activities of Deutsche Telekom on Interactive Video Services with Focus on Multimedia Server Design", Proceedings of the 3rd International Conference on Communicating by Image and Multimedia (IMAGE'COM) 1997, Bordeaux, France.

[3] Gross, B., Breide, S., Ruppel, W.: „Services-on-Demand Laboratory Demonstrator with PC-based Set-Top-Boxes", Proceedings of the International Workshop on HDTV, Los Angeles, October 1996

[4] Achtmann, K., Döring, K.-H., Herber R., Komp, G: „An ATM-based Demonstration Model for Multimedia Services Using Different Access Netwworks" published in „Multimedia Applications, Services and Techniques - ECMAST 1997" by Springer Verlag, Berlin 1997

[5] Digital Audio-Visual Council: DAVIC 1.0 Specification, December 1995, Geneva, Switzerland.

[6] Digital Audio-Visual Council: DAVIC 1.1 Specification, September 1996, Geneva, Switzerland.

[7] Digital Audio-Visual Council: DAVIC 1.2 Specification, December 1996, Geneva, Switzerland.

[8] Digital Audio-Visual Council: DAVIC 1.3 Specification, September 1997, Geneva, Switzerland.

[9] Balabanian, V., Casey, L., Greene, N.: „*An Introduction to Digital Storage Media - Command and Control*", http://drogo.cselt.stet.it/ufv/leonardo/mpeg/documents/dsmcc.htm

[10] ISO/IEC International Standard 13818-6: „*MPEG-2 Digital Storage Media Command & Control*".

[11] Deutsche Telekom: „*KIT - Windows-based Kernel for Intelligent Communication Terminals*", KIT Version 1.2, April 1996.

[12] European Telecommunications Standards Institute: „*Enhanced Men Machine Interface for Videotex and Multimedia / Hypermedia Information Retrieval Services (VEMMI)*", prETS 300 709

[13] Olivetti & Oracle Research Laboratory: http://www.orl.co.uk/omniORB

[14] Olivetti & Oracle Research Laboratory: http://www.orl.co.uk/omniORB/omniORBPerformance.html

Resource Selection
in Heterogeneous Communication Environments
Using the Teleservice Descriptor

Tom Pfeifer, Stefan Arbanowski, Radu Popescu-Zeletin

Technical University of Berlin
Open Communication Systems (OKS)
Franklinstr. 28, 10587 Berlin, Germany
e-mail: pfeifer@fokus.gmd.de

Abstract: Automated processes in distributed communication environments require tools for unifying heterogeneous multimedia services. The Teleservice Descriptor is introduced for generic handling and integration of traditional and innovative forms of communication. The Intelligent Resource Selector applies this descriptor for dynamic selection of communication end points and combination of necessary converters for service interworking. The Intelligent Personal Communication Support System provides the test-bed for the implementation of the developed algorithms, applicable in CPE, TINA and IN solutions.

Keywords: Multimedia communication in distributed, heterogeneous networking and communication environments, Media Conversion, Quality of Service, Personal Communication Support, Personal Mobility.

1. Introduction

Future communication, as described in the Virtual Home Environment (VHE) concept [2] within the emerging Universal Mobile Telecommunication System (UMTS) standards [1], aims to deliver to deliver *"information any time, any place, in any form"*. Systems have been introduced to manage and to control the global reachability of people in order to maximise or to filter it – independent of their location, the used communication medium, and the applied human communication interaction (asynchronous or synchronous). Additionally, these systems are to provide access to asynchronous messages from everywhere and with any kind of terminal.

In this context, an overview of the *intelligent Personal Communication Support System* (iPCSS) has been presented at the previous workshop [26], supporting the three important aspects of Personal Communication Support, namely *Personal Mobility*, *Service Personalization*, and *Service Interoperability* in distributed multimedia environments. For the latter case, capabilities are required that enable dynamic/intelligent content handling and conversion of different media types and media formats in order to deliver information in any form.

The term 'intelligence' refers to the capability of the iPCSS to make certain decisions within user-defined limits by itself, therefore relieving the user from pre-planning every possible situation in his communication environment.

The research is backed by practical implementations of the system, performed jointly by TU Berlin, GMD Research Center for Open Communication Systems (FOKUS), and Deutsche Telekom Berkom.

While the previous paper provided an overview of conversion capabilities and the complexity of the system, we focus now on the technical aspect of unifying access to heterogeneous multimedia services, comprising and integrating traditional and innovative forms of communication.

Section 2 analyses the respective requirements, while section 3 introduces the Tele-service Descriptor for generic handling of telecommunication services. Section 4 presents the Intelligent Resource Selector, applying this descriptor for the dynamic selection of communication end points and the combination of necessary information converters for service interworking. The concept and experiences of the implementation within the iPCSS are sketched in section 5. The outlook points to possible applications of the core system within a wide scale of scenarios, comprising CPE, TINA [7] and Intelligent Network (IN) [19] solutions.

2. Heterogeneous Multimedia Communication Facilities

Humans communicate with each other in many very different ways. Transport media as well as presentation media of the different ways of communication have individual properties and heterogeneous characteristics.

Electronic, digitized and computerized media handling has unified a lot of aspects in terms of transport and storage of communication data, however it has brought up completely different, incompatible solutions for each task, which are just recently going to be bridged.

A computerized communication system dedicated to a specific medium, such as a telephone system with digital switching, has been built to take care by default of all specific properties of this very form of communication.

In the next step, a system dedicated to bridge two forms of communication, such as a system dedicated to forward e-mail by voice telephone, can handle the specific properties of these two forms of communication in a predefined, pre-planned and well adjusted manner.

As our current research is focused on automated, intelligent decisions how to handle actual communication request most appropriate, and therefore to bridge any form of communication with any other, we face the problem to handle the full variety of individual characteristics for each event of communication.

For automated handling, the heterogeneity of communication has to be classified precisely, allowing the system to match components for various purposes within a huge construction kit. Elements required for the adaptation and conversion of communication media, such as connecting gateways, the selection, configuration and building of converter chains, have to be in the focus of such classification, providing a generic description of the semantic carried by each of the media used.

While our implementation test-bed uses a CPE/CPN environment, the underlying technology needs to be scalable for integration in future global communication systems, considering the recent developments in TINA [8] and Intelligent Networks (IN) [19].

3. Generic Approach: The Teleservice Descriptor (TSD)

The basics for global connectivity and ubiquitous computing are at first the hardware supporting all the necessary performance and the broadband networks for high speed multimedia services, and at second the software controlling all services. This software has to handle a huge set of telecommunication and information services, which makes it impossible to create a controlling software for each service. To design a system, which is able to handle the whole set of services and combine services to

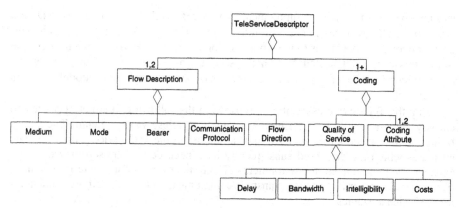

Fig. 1. **Structure of the Teleservice Descriptor**

enable service interworking (e.g. a GSM user can be invited to an ATM-based video-conference) a unique description for all services is required.

This section describes the different parameters for such a generic description and suggests a complete set of attributes for a generic service descriptor. It leads to the structure presented in Figure 1.

To prevent misunderstanding with other meanings of "service", we introduce the term *Teleservice*, describing the special meaning in the context of telecommunication.

attribute	open lists of possible values
Medium	image, video, speech, audio, video & audio, text, file
Mode	synchronous, asynchronous
Flow Direction	sink, source, sink-source
Bearer	ISDN, B-ISDN, ATM, Ethernet, FDDI
Communication Protocol	X.400, MMC, H.320, FTP, RFC822, MMM, HTTP, SNMP, G3, G4
Coding	GIF, JPEG, BMP, TIFF, G3, G4, XPM, BPM, PGM, PPM, PNM, XBM, XWD, MPEG1-Video, MPEG2-Video, AVI, MOV, FLIC, MJPEG, H.261, MPEG1-Video & Audio, MPEG2-Video & Audio, ADPCM, LPC, PCM, MLAW, ALAW, AU, WAV, VOC, SND, MIDI, S3M, 669, MPEG-Audio
Quality of Service	bandwidth, delay, costs, intelligibility

Table 1 **Teleservice Descriptor Attributes**

Telecommunication services enable users to communicate in different ways using these services. The specific characteristic of a dedicated Teleservice can be described with a set of attributes, which has been defined to describe the meaning of the term "service" within this document. These attributes are in detail "medium", "mode", "information flow", "bearer", "communication protocol", "coding", and "quality of service" (see examples in Table 1). Only the complete set of the attributes describes a Teleservice. The entries of *possible values* given in the table represent incomplete, open lists, which lists have to be updated day by day with new bearers, communication protocols, and codings.

Other approaches have been influenced by the different kinds of communication. In this context the term service has been used for any kind of information exchange. The meaning of service has been overloaded frequently. Different network technologies classify their services to improve the handling of terminology. Each approach uses different classifications and different terminologies (e.g. Intelligent Networks (IN) services may be split into basic or supplementary services, while Integrated Service Digital Network (ISDN) services can be split into bearer services, Teleservices or supplementary services). TINA-C aims to be applicable for all types of services, including simple bearer services, sophisticated multimedia services, management services, and operation services. The benefit is, that for each class of service a reference model has been provided [8].

Using the Teleservice description described in the following, it is not necessary to classify Teleservices or to describe the meaning of Teleservices in detail. The values fitting the introduced attributes have to describe the whole Teleservice properties. The attributes which are discussed subsequently have been defined by scrutinizing some existing telecommunication environments. Three main types of attribute groups have been derived – general attributes, transport/communication related attributes, and quality of service attributes.

3.1. General Attributes

At first, three Teleservice attributes will be introduced which are derived from information exchange of human beings. The attributes medium, mode, and information flow have no technical background. If somebody wants to describe only the communication between two human beings, without any technical equipment, these attributes would be sufficient to characterize the transmission aspect of the information exchange.

3.1.1. Medium

From the broad scope of meanings of the term "medium", comprising perception media, representation media, presentation media, storage media, transmission media, and information exchange media [17], this paper considers the aspects defined for computerized multimedia communication, in particular the technical media stimulating the human senses for information exchange, namely perception media, such as text, graphic, picture, speech, sound, and music.

For building the TSD, the defined values of the attribute *medium* are shown in Table 2.

values	short description
image	is a still image
video	is a motion picture
audio	is any kind of sound
speech	is a single human or computer generated (speech synthesis) voice
video & audio	is video with simultaneous audio
text	is text coded in any format
file	is not further interpreted, arbitrary data

Table 2   **Values for the attribute Medium**

Some of the values refer to the same human sense, other senses are omitted. While video and still image both stimulate the visual sense, the impact for communication systems (data volume, semantic, cost) as well as the cognitive processes involved are completely different. Audio and speech (as the audible representation of a natural language), as defined above, are useful distinguishments for the purpose of automatic handling these media within an intelligent communication environment, and so is the combined class of video&audio.

Of course, the list could be easily extended if experimental approaches to stimulate the other human senses like touch (Braille devices, data gloves), smell or taste should be integrated into the respective scenario.

On the other hand, communication may involve data not addressed towards the human senses but sent to a special user application (e.g. large tables of stock prices). Such data are covered by the value 'file'. These data should not be altered or interpreted by intermediate communication systems.

Conversion

In the previous section, the term "conversion" was implicitly used. Now it should be discussed from a pragmatic point of view, which media could be converted into another and what the result of such a conversion could be expected. A short summary is given in Table 3. The media types in the left column represent the source media. On the top of the table the target medium is specified. The table shows the results of the conversion from source to target.

Not in all cases the semantic of the source media survive the conversion. It is then only possible to produce a message that an entity from the source media type was detected. An example is the conversion from image to speech. Despite research in image recognition, it is nearly impossible to convert from image to speech preserving the full semantic of the source image. A solution is to notify the receiver with a synthetically created speech message: "There is an image for you", or "Image containing to faces".

Some conversions are only possible by using more then one conversion steps, for instance speech to image. Speech can be recognized by special software packages, analyzing the language and later the spoken words. The type of the result may be 'text'. Finally, we convert this text to image by pixel or postscript representations.

It is not possible in each case to convert without loss of semantic. When converting video&audio to audio, the video part is lost. While converting audio to video&audio, the video part of video&audio is blank or replaced by an still image.

	image	video	audio	speech	video & audio	text	file
image	trivial format conversion	still video	message	message	still video	message	save image as file
video	still picture – selected frame	trivial format conversion	message	message	silent video	message	save video as file
audio	visualization in different kinds	music notes on video	trivial format conversion	message	"black" video	message	save audio as file
speech	characters on image	characters on video	yes	trivial format conversion	"black" video	speech recognition	save speech as file
video & audio	combine video & audio to image	lost audio, or display of music notes	lost video	message	trivial format conversion	message	save video & audio as file
text	characters on image	characters on video	speech synthesis	speech synthesis	text scroll; speech synthesis	trivial format conversion	save text as file
file	message	message	message	message	message	message	trivial form. conversion

Table 3   Possible Media Conversions

All entries in Table 3 titled "trivial" are not difficult to convert, using state of the art conversion mechanisms. The column "file" is also easy to treat. Without any problems, it is possible to represent all media types as file.

3.1.2. Mode

The mode of a Teleservice can be synchronous or asynchronous (Table 4). Asynchronous Teleservices do not depend on any time conditions, therefore best effort strategies can be applied. E.g. the store and forward message handling within the internet works on these principles. An example for a connection oriented asynchronous Teleservice is a sophisticated automatic answering machine which forwards the recorded message to a certain telephone number.

values	short description
synchronous	both parties involved at a time
asynchronous	one party involved at a time

Table 4   **Values for the Attribute Mode**

Synchronous Teleservices are real-time dependent. A fixed time frame exists for transmitting a part of the information from the sender to the receiver (end-to-end delay). If the time frame is exceeded the synchronous communication will be broken. A subclass of synchronous communication is isochronous communication, whith constraints to the jitter.

3.1.3. Flow Direction

Using this attribute (Table 5), the direction of the information exchange can be determined. The "flow direction" of a Teleservice can also be described as duplex or simplex, or as unidirectional and bi-directional. Unfortunately these values do not express anything about the fact which side of the connection receives and which sends information, which can be completely described with the terms source and sink.

values	short description
sink	the Teleservice receives information (unidirectional, simplex)
source	the Teleservice transmits information (unidirectional, simplex)
sink/source	the Teleservice receives and transmits information (bi-directional, duplex)

Table 5   **Values for the Attribute Information Flow**

3.2. Transport and Communication Related Attributes

In contrast to the general attributes above, the transport and communication related attributes have a technical background. With these attributes it is possible to describe the technical characteristic of a Teleservice in a generic way. Therefore, it was necessary to abstract from specific and proprietary parameterization of the underlying technologies. An approach for mapping a Teleservice to the parameterization of a physical resource leads to a system of dynamic resource configuration management, which is part of our research, but not covered by this paper.

3.2.1. Bearer (CPE-Bearer)

The "bearer" attribute describes the physical network technology on which the Teleservice will be transported to a service gateway which can be controlled by the intelligent communication environment (Table 6). These gateways enable the connection to different network technologies outside the Customer Premises Equipment (CPE) / Customer Premises Network (CPN). For instance, the connection between an old Public Switched Telephone Network (PSTN) and a CPE/CPN is only possible via service gateways. The service gateway can perform a bearer conversion (e.g. from ISDN to ATM). The knowledge about the underlying network technology is necessary to address the respective gateway.

possible values	short description
ATM	Asynchronous Transfer Mode (ITU-T I.361)
FDDI	Fibre Distributed Data Interface (ITU-T 3914x)
ISDN	Integrated Service Digital Network (ITU-T I.320)
B-ISDN	Broadband Integrated Service Digital Network (ITU-T I.321)
DQDB	Distributed Queue Dual Bus (IEEE 802.6)
Ethernet	normal 10Mbit Ethernet (IEEE 802.3)
GSM	Global System for Mobile Communication
DCS-1800	Digital Cellular System
PSTN	Public Switched Telephone Network

Table 6   Possible Values for the Attribute Bearer [27]

The starting point for this approach is the existence of distributed communication environments, which are based on distributed heterogeneous networks. In such an environment, the underlying network characteristic is transparent to the users. Different network technologies are interconnected. Solutions for interconnecting different kind of network technologies exist and have been tested for a long time. It is possible to address applications directly, independent of the underlying network. An example for addressing an application is File Transfer Protocol (FTP). Although the user addressing an application does not know where the host computer is located and what the kind of network the information is transported through. It could be a TCP/IP connection over Ethernet or over ATM. For using services inside of the CPE/CPN the attribute bearer is useless, but for the description of services transmitted over service gateways (see above), the attribute "bearer" is required.

3.2.2. Communication protocol

The "communication protocol" (Table 7) is an attribute holding information about the protocol which is used between the application related entities of the communication end-points. The possible values have been designed as an open list, allowing the addition of new communication protocols to the list.

In the context of the attribute "bearer" above the fact was discussed that in customer premises equipment it is possible to address user applications. To communicate with these applications it is necessary to define the communication protocol. On the basis of this protocol, the user applications can exchange information (e.g. a video conference is based on the H.320 standard).

Depending on the usage of the TSD it might be necessary in some scenarios to define a further attribute "Transport protocol", covering such lower level protocols as TCP/IP, X.25, IPv6, ISDN, RTTP, ZMODEM. However, as the systems developed within our projects employ middleware platforms like CORBA in distributed communication environments, these protocols are transparent.

example value	short description
X.400	Message Handling System (ITU-T X.400)
RFC822-mail	Internet Mail (RFC 822)
SNMP	Simple Network Management Protocol (RFC 1270)
MMM	multimedia mail
H.261	video compression for p x 64 kBit/s (ITU-T H.261)
H.320	format for narrow-band visual telephone services (ITU-T H.320)
HTTP	Hypertext Transport Protocol
ISDN	Integrated Service Digital Network (ITU-T I.320)
FTP	File Transfer Protocol (RFC 765)
DAP	Directory Access Protocol (ITU-T X.500)

Table 7   Possible Values for the Attribute Communication Protocol [27]

3.2.3. Coding

The attribute "coding" (Table 8) describes the format the data is stored or transmitted in the system.

A format is a guideline how data have to be structured. It divides the data in control data (meta data) and usage data. The control data is mostly stored in a header and defines the areas of usage data. For the usage data the order of information and possible used compression algorithms are stored. Data streams without a header contain the

possible values	related medium
GIF, JPEG, BMP, TIFF, G3/G4 Fax	Image
MPEG1, MPEG2, AVI, MOV, FLIC	Video / Video & Audio
ADPMC, LPC, PCM, G.711, ALAW, MLAW	Audio / Speech
AU, WAV, VOC, SND, MIDI, S3M, MPEG	Audio
Binaries and other non-interpreted formats or raw data	File
ASCII, ISO 8859, EBCDIC	Text
HTML, SGML, LaTeX, PDF, PS, DOC, FM	File
ARJ, GZIP, ZIP, LHA, MIME, uuencode, Base 64	File

Table 8   Possible Values for the Attribute Coding [18, 27]

control information between the usage information or lack any control information. In this case, the format must be well defined and fixed. These data formats could not only be stored in a file system, but appear also in a continuous data flow coming from a stream device like a video-camera. "Coding" is sorted by the different existing file types. Therefore the used names for the attributes come from the corresponding file type. In due to the endless number of existing codings this attribute should be designed as a growing list of supported codings.

3.3. Quality of Service Attributes

To fulfil the vision for future telecommunication environments to deliver "information in any time, any place, in any form", an unlimited spectrum of telecommunication services will be offered by different service providers. The spectrum of conceivable services ranges from simple communication services up to complex distributed multimedia services. To enable "information in any form" requires to convert certain communication media into another media or at least into another format of the same medium, leading to a support of higher flexibility of terminals or applications.

Such conversions are currently done by stand-alone processes, realized in software or designed as hardware. Examples for conversion processes in this context are Text-to-Speech (TTS), Optical Character Recognition (OCR), and Speech Recognition. Future telecommunication environments need to be able to combine such processes of conversion to enable service interworking in a generic way. Arbitrary combinations of converters require unified converter interfaces.

Providing such conversions in an integrated framework with well defined interfaces, able to combine any conceivable combination of converters to chains (e.g. fax -> image conversion -> OCR -> text conversion -> TTS -> audio format conversion), leads into the problem of evaluating the quality of the outcome of each conversion in the chain. The assessment of that quality enables the finding of a most appropriate converter chain and a most appropriate terminal, applicable for the requested service. This section introduces Quality of Service attributes for this specific purpose. [7]

A number of Quality of Service attributes have been introduced in [26], some of them are discussed below. However, not all of them could be used for the design. For enabling an easy-to-handle algorithm which is able to compare two complete sets of Quality of Service attributes it is necessary to reduce the number of attributes. The attributes proposed for usage are "bandwidth", "delay", "cost" and "intelligibility". Please note that the term QoS has been adopted from the networking context, but is used in a different way with other parameter sets here.

3.3.1. Bandwidth

The parameter "bandwidth" (Table 9) describes the required transmission resources. The attribute contains information about the minimum bandwidth which is needed for the Teleservice. The bandwidth is the resulting data rate during the connection. It is suggested to use as values numbers in Mbit/s. There is also a need to set up connections with no dedicated bandwidth. These connections can be initialized with 0 Mbit/s as a special value, e.g. e-mail or G3-fax.

possible values	short description
150 Mbit/sec	high speed connection - Constant Bitrate transmission
14400 baud	small-band modem connection - Constant Bitrate transmission
20 - 50 Mbit/sec	Variable Bitrate transmission

Table 9   **Possible Values for the Attribute Bandwidth**

3.3.2. Delay

The term "delay" (Table 10) is used in this document synonymously to the term "end-to-end delay". It describes the time the information exchange needs from the transmitter to the receiver (the receiver is not the communication endpoint, but the human being who uses the communication endpoint). An alternative name could be "global delay".

possible value	short description
1 ms	describes a definite time value

Table 10   **Possible Value for the Attribute Delay**

"Delay" is composed of the three sub-types network transmission delay, computing delay, and buffering delay. The network transmission delay is caused by the limited speed of signals and influenced by all factors which slow down the connection (e.g.

overload, failure, and breakdowns). A computing delay results from any calculation process (e.g. conversion processes). Buffering is required in flow control related mechanism.

3.3.3. Cost

The parameter "cost" (Table 11) refers to all use of computational resources as well as to the transmission cost. In a distributed communication environment it is possible to abstract from the used network and to use different computational resources during a communication. Different converters and other service supporting resources could be used transparently to the user. For charging and billing each provider can define a certain amount of concrete resource which a user has to pay for the usage of it. To calculate how much a user has to pay for a concrete communication, every value has to be summed up to the total amount.

possible values	short description
US$ 0,03	a certain amount of a certain currency
3 units	a certain amount of a virtual currency

Table 11   Possible Values for the Attribute Cost

In the area of computing the term "cost" is also used to describe the required resources for a computing process. In face of that, the next attribute is modelled. It could be seen as a subclass of the attribute "cost" (due to the various aspects to cover with the attribute, it is in some places used in its plural form "costs").

A problem arises when time dependent rates and flat rates have to be compared within the cost estimation in a QoS evaluation. A possible solution would be to evaluate the time dependent rates for average communication time.

3.3.4. Intelligibility

The parameter "Intelligibility" is the most important determiner for the correct transport of the semantic of the information during a Teleservice conversion process. Defined in Webster's dictionary as "capability of being understood or comprehended", it describes whether the human being perceiving the output of the conversion process is able to recognize its semantic correctly or not. The term is mostly used in the context of complex conversion, as text to speech, optical character recognition and speech recognition.

Concrete values for this attribute are difficult to determine (Table 12). Generally, such assessments are only possible by human beings. Computer supported fax processing (i.e. the received fax will be converted into text and then into speech) is an example for a need of the determination of "intelligibility". The problem using a computer is to asses the received fax, whether the semantic is lost or not. A possible solution for this specific problem is counting spelling errors after OCR: If the number of recognized words is to small, the received fax is useless for the conversion into speech. The attribute "intelligibility" can be divides in various subordinated categories, such as

- Error Probability, which is the more technical version of the intelligibility
- Quality degradation due to (accumulated) lossy compression
- Quality degradation due to entropy reduction (color reduction, quantification, scaling, resampling).

These categories deliver ways to measure and determine values for the intelligibility. Technical categories can evaluate bit error rates, mean square differences, noise

possible values	short description
hard to decipher	value oriented on human beings
50%	semantic loss after conversion

Table 12   Possible Values for the Attribute Intelligibility

levels, while on a language level e.g. the hit counts of spelling checkers determines the kind of the language as well as the usability of the outcome of the OCR.

3.3.5. The Attribute Quality of Service

For a first approach it is impossible to consider the whole complexity of possible Quality of Service attributes. A useful selection has been made above with bandwidth, delay, costs, and intelligibility (Table 13). This is a strong simplification, but it is useful for demonstrating a first functional model. All usage of resources is hereby covered by the parameter cost, and all errors, compression losses, degradation influence the intelligibility.

possible values	short description
bandwidth	required transmission resources
delay	temporal stoppage of the communication
cost	usage charge of any equipment / resources
intelligibility	capability of being understood or comprehended

Table 13   The Attribute Quality of Service

3.4. Summary

Employing the developed Teleservice Descriptor, it is possible to describe all Teleservices in a generic way. For the demonstration of this ability, six example cases of communication are shown in Table 14.

The first example represents the easiest case – the conventional ISDN telephony. The used "communication protocol" is only G.711, because the signalling before the connection is enabled (via D-channel protocol Q.931) [27] is irrelevant. The "intelligibility" should be set to 100%, if only connections between ISDN telephones are desired (the process of digitizing the natural speech through the microphone in the ISDN-telephone is out of the scope of this study).

The asterisks in the table have the meaning of a wildcard. It is possible to use all possible values in the asterisk context, or the asterisked attribute is useless in the special context. For instance, for the Teleservice *chat* many chat protocols are conceivable and the Quality of Service attributes do not make any difference. The differences between the values of "cost" for *ISDN telephony*, *File Transfer* and *WWW via Cellular* are the semantic of the values. ISDN telephony means the costs per unit, *File Transfer* the costs per megabyte, and *WWW via Cellular* the costs per minute.

The intelligibility of the ISDN-based Video Conference is 70 percent. This means, that the quality reduction through the video compression is to high to keep the original contents.

Based on concrete Teleservice as described above, it is now possible to describe terminal capabilities and conversion processes in a generic way for enabling intelligent selection algorithms. Beside this, it is necessary to design a model for creating the concrete values for a Teleservice automatically, in particular the Quality of Service attributes.

Service Expl. Attributes	ISDN telephony	ASCII Chat	File Transfer	WWW via Cellular	Joined Editing of Documents	Video Conference
Medium	speech	text	file	*	*	video&audio
Mode	synchronous	synchronous	asynchronous	asynchronous	asynchronous	synchronous
Flow Direct.	sink/source	sink/source	source (sink)	source (sink)	sink/source	sink/source
Bearer	ISDN	*	ATM	DCS-1800	*	ISDN
Comm. Prot.	ISDN	*	FTP	HTTP	CSCW	H.320
Coding	G.711	ASCII	file	ASCII	file	*
QoS-Param.						
Bandwidth	64kbit/s	*	25Mbit/s	9600baud/s	64kbit-2Mbit/s	128kbit/s
Delay	0	*	*	1 sec	*	0,3 sec
Cost	DM 0.12/unit	*	$15/Mb	DM1.80/min	$ 0/unit	DM 3.60/min
Intelligibility	100%	*	*	100%	100%	70%

Table 14   Examples for Teleservice Descriptor Usage

4. Automatic Resource Selection

This section is based on an approach for a *Teleservice Descriptor* developed above, describing each kind of Teleservice in a generic way. In an intelligent communication environment it is not sufficient to use such a description for services only. Beside the services, a generic *terminal description* and *converter description* is required for the construction of a selection algorithm to find the most appropriate terminal for a requested Teleservice at the user's current location dynamically. The selection can be divided into several parts: selecting only terminals, or selecting multiple converters and one terminal to build a converter chain. Within the selection process, the TSD is applied to various objects involved, as shown for a less complex case in Figure 2. The details will be explained within this section.

4.1. Representations of Physical Resources

The following analysis is based on the assumption that an end-user system is able to handle a Teleservice. Additionally it is necessary to assign Teleservice Descriptor attributes to terminals and converters, so that the input and/or output of a terminal or converter could be described in a generic way. Media conversion mechanisms can be done by converters implemented in software or designed as hardware. These converters need also to be described with the Teleservice Descriptor attributes. The concrete assignment of the attributes is different for each terminal. The result of the assignments is a generic representation for terminals and converters, which is also used for the selection mechanisms.

4.1.1. Service Access Point

A Service Access Point (SAP) is a logical representation for a physical communication endpoint which is able to handle a specific Teleservice. The underlying physical resource could be hardware (e.g. a telephone) or software (e.g. a software package for video conferencing). It describes the communication capabilities of this physical

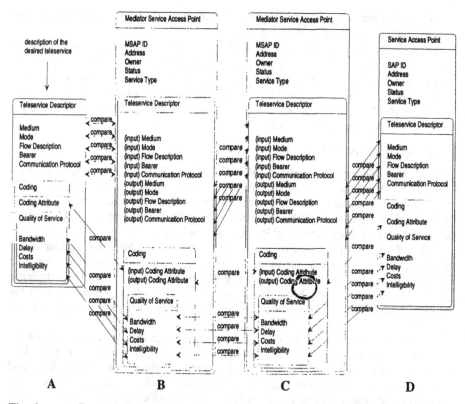

Fig. 2.   **Dynamic Resource Selection: Converter Chain containing TSD, MSAP, SAP**

resource, i.e. which Teleservice the resource supports. This description is done by the complete set of Teleservice Descriptor attributes.

A complete set of TSD attributes is used. Additional attributes for a complete terminal description are "SAP ID", "Address", "Owner", "Status", "Servicetype". The complete Service Access Point model is shown in part D of Figure 2.

Due to physical resources which support more than one Teleservice at a time, it is conceivable that a Service Access Point has to contain more then one TSDs. This approach can be avoided by representing such a resource in the respective number of SAPs. The only exception is that a Service Access Point can contain as many Coding and QoS attributes as it supports different codings. The two attributes have to be seen as a n-tuple, because the QoS of a coding describes the quality degradation of the Teleservice during the processing by the terminal (e.g. the representation of a 24 bit image on an 8 bit screen). The n-tuple of the "Coding" attribute and the "Quality of Service" attribute is stored in the attribute "Coding". Most software SAPs support many codings (e.g. an image viewer supports many formats of image).

The attribute "SAP ID" guarantees a unique name for each Service Access Point for access / administration activities. In the attribute "Address" the physical address of the terminal is stored (i.e. an ISDN-telephone number, an IP-address, a hostname, or a port on a distinguished host). Sometimes it is useful to assign a Service Access Point to a certain user of the system or to assign groups of Service Access Points to an organizational unit for simplifying administration. The attribute "Owner" stores the information to which user a terminal is assigned to.

The attribute "Servicetype" can be used for storing a human readable description (e.g. "ISDN based video-conferencing"). The attribute "Status" mainly stores the information whether a resource is usable or not; values are "busy" (a terminal is currently working), "idle" (a terminal is currently unused), "up" (the terminal is not busy, but can nevertheless not be used – "it is coming up"), and "down" (an administrator has disabled the usage of a terminal).

4.1.2. Mediator Service Access Point

A Mediator Service Access Point (MSAP) is a logical representation for a physical service converter which is able to handle a specific Teleservice on its input, to convert this Teleservice into another Teleservice, and to pass the result to its output. The underlying physical resource could be hardware (e.g. a MPEG real-time encoder board) or software (e.g. an Optical Character Recognition package).

The MSAP (see the models in part B and C of Figure 2) describes the conversion capabilities of this physical resource. In opposite to the SAP described above, an MSAP contains an input and an output. Both are described by the TSD attributes Media, Mode, Information Flow, Bearer, and Communication Protocol, which are therefore required twice. ID, Address, Owner, Status and Service type are equivalent to the SAP.

The major difference to the SAP results from the fact that a converter in general reduces the Quality of Service and produces a new Teleservice on its output. This reduction is described with one "Quality of Service" attribute for an MSAP. Input and output do not have a separate "Qualities of Service" each, because the semantic of "Quality of Service" is different to the SAP – it describes the Quality of Service related conversion behaviour of a MSAP. In this case, "Quality of Service" describes the quality reduction for two specific "Coding attributes" (input coding attribute, output coding attribute) through the converter and is ternary assigned to two "Coding Attributes" (for ternary associations see [13]).

The input coding attribute is converted to the output coding attribute with the Quality of Service of the MSAP. The triple of the two "Coding attributes" and the one "Quality of Service" attribute is stored in the attribute "Coding".

A Mediator Service Access Point allows to specify more the one "Coding" attribute like a Service Access Point for the same reasons than described in the previous section. Also an MSAP can support several codings and particular conversions. The additional attributes in an MSAP are the same than in a SAP.

4.1.3. Virtual Access Point

The Virtual Access Point (VAP) represents a service generic or service neutral communication end-point, to provide independence of any existing, real telecommunication service or device. A VAP has a fixed relationship to a specific organizational unit (e.g. an office, or a desktop in an open-plan office) and can be understood as a collection of communication capabilities, which are represented as Service Access Points. (Figure 3a)

4.1.4. Intelligent Resource Selector

One of the most difficult problems is the dynamic resource selection. The Intelligent Resource Selector (IReS) provides an algorithm for the dynamical finding of the

Fig. 3.   **a) Virtual Access Point**   **b) Intelligent Resource Selector**

most appropriate converter chain related to a requested Teleservice if no terminal can support it without conversion. The IReS is parameterized with

- the list of Service Access Points contained in the respective VAP,
- additional Service Access Points (terminal registration),
- forbidden Service Access Points (the user does not want to use),
- preferred Service Access Points (terminals the user prefers),
- and possibly with any other user preferences the Intelligent Resource Selector can take into account.

An Intelligent Resource Selector itself maintains a list of all available Mediator Service Access Points. (Figure 3b).

4.2. Selection Processes

The *dynamic resource selection* is divided into two parts, the *single* and the *multiple resource selection* (Figure 4). The first is a selection between terminals described and represented as SAPs, whereas the latter works on SAPs and MSAPs to build converter chains.

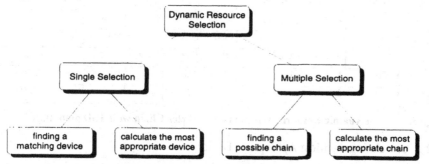

Fig. 4.   **Dynamic Resource Selection - Different Selection Processes**

A converter chain (see example in Figure 5) consists of an in-gateway as the input interface, several converters, an optional out-gateway, and one terminal, representing the communication endpoint. A converter chain can contain as much converters as there are available. The converters connected in series have only to meet one condition: The Teleservice produced by the output of a connected converter has to match the Teleservice of the input of its successor.

The *single resource selection* can be further divided in *finding a single matching device* and *calculating the most appropriate device* (SAP). The most appropriate terminal for a Teleservice is that terminal which supports the desired Teleservice with the best QoS and satisfies all the preferences the user has been made.

The *multiple resource selection* can be further divided in *finding a possible chain* and *calculate the most appropriate chain*. The namely introduced selection processes are strongly different in calculating expenditure. *Calculate the most appropriate chain* is the selection process with the highest calculating expenditure.

4.2.1. Finding a single Matching Device

Finding a matching terminal is very simple. Every TSD in all of the SAPs (contained in the very Virtual Access Point the user was located) has to be compared with the Teleservice Descriptor in the parameterization of the process. The first Service Access Point which has the same TSD as the desired Teleservice is used as the result. Other Service Access Points which could also match the requested Teleservice will be ignored. The algorithm follows the "first fit"-strategy.

Two TSDs match if each pair of attributes has corresponding values. It means that the attributes "Medium", "Mode", "Flow Description", "Bearer", and "Communication Protocol" must be the same, while the QoS attributes "Bandwidth", "Delay", "Cost", and "Intelligibility" in the SAP have to be the same or better values as the desired Teleservice (e.g. if a "bandwidth" of 5 Mbit/s is required a SAP with a "bandwidth" of 10 Mbit/s could be selected). The QoS attribute of the requested Teleservice is the minimum QoS for the available terminals.

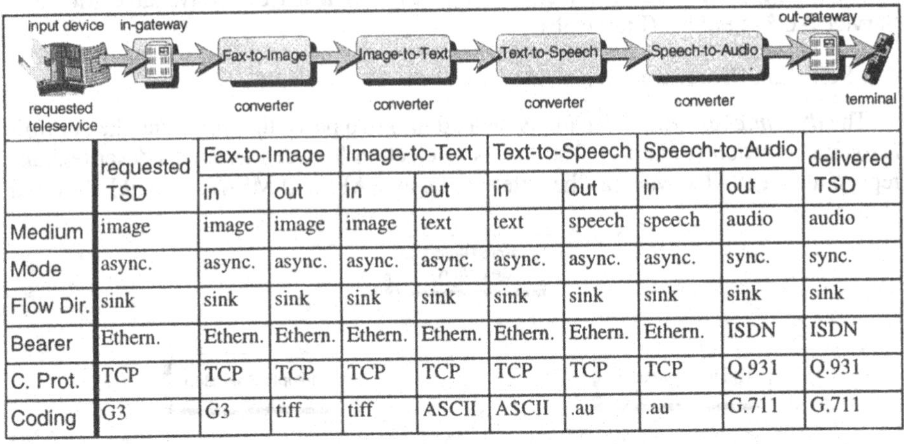

	requested TSD	Fax-to-Image		Image-to-Text		Text-to-Speech		Speech-to-Audio		delivered TSD
		in	out	in	out	in	out	in	out	
Medium	image	image	image	image	text	text	speech	speech	audio	audio
Mode	async.	async.	async.	async.	async.	async.	async.	async.	sync.	sync.
Flow Dir.	sink	sink	sink	sink	sink	sink	sink	sink	sink	sink
Bearer	Ethern.	Ethern.	Ethern.	Ethern.	Ethern.	Ethern.	Ethern.	Ethern.	ISDN	ISDN
C. Prot.	TCP	TCP	TCP	TCP	TCP	TCP	TCP	TCP	Q.931	Q.931
Coding	G3	G3	tiff	tiff	ASCII	ASCII	.au	.au	G.711	G.711

Fig. 5.   Dynamic Resource Selection – Converter Chain with TSD mappings

4.2.2. Calculating the Most Appropriate Device

Calculating the most appropriate device is based on *finding a matching device*. All SAPs within the very Virtual Access Point are compared with the desired TSD. If there is more than one matching terminal, it depends on user predefined specifications which one has to be selected (terminal owner, preferred SAP etc.).

Based on the additional parameters and the requested Teleservice, the IReS chooses the most appropriate SAP. Therefore the QoS values of all possible SAPs are calculated. The QoS comparison is not trivial. The attributes "bandwidth", "delay", "costs", and "intelligibility" have to be weighed for making a solution possible.

A preliminary evaluation model is the selection of the SAP which has

- the highest "Intelligibility",
- the lowest "costs",
- the shortest "delay", and
- the bandwidth corresponding to the requested Teleservice.

The problem is to define the priority of attributes, which may also depend on user preferences. One user communicates images for publication and insists on the highest intelligibility, the other one is short of money and satisfied reading nearly white noise. A discussion of selection strategies can be found in [25].

4.2.3. Finding a Possible Converter Chain

The *multiple resource selection* is an enhancement of the *single resource selection*. Additional mediator units are involved in the selection process and the result of a *multiple resource selection* could be a combination of many mediator units and one terminal (forming a converter chain – see above). A multiple resource selection is only necessary in the case that no matching SAP was found in the *single resource selection*.

Calculating a converter chain needs a complex algorithm. Each possible combination of converters with a final terminal must be evaluated. The theoretical space of search is tremendous. Figure 6 shows the possible space of search represented as a tree, giving us the impression how the number of possible combination grows with the number of available terminals and converters. The figure demonstrates a scenario with three converters and four terminals, leading into 60 possible converter chains, one of them emphasized. Five converters and five terminals allow 600 combinations, and one hundred converters and terminals lead to over 500.000 theoretical possibilities. Realistic values for the number of terminals and converters are substantially higher, because all available software tools supporting a specific Teleservice have to be represented as a SAP, so that the number of possible converter chains could be millions.

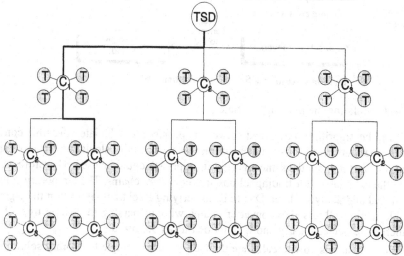

Fig. 6.   Selection of possible Converter Chains (C={C$_1$ ... C$_3$} T={T$_1$ ... T$_4$})

The main task of the algorithm is therefore to reduce the space of search. Traversing of trees is a well-known method in the field of computer science [16]. Different procedures such as Backtracking or Branch-and-Bound are applicable. The chosen approach is based on a Backtracking-algorithm. This algorithm traverses a tree first down to a leave and tries then to find the next leave. For finding one possible chain (a valid path through the tree down to a leave) this is the best algorithm, because the number of temporal solutions and iterations is much smaller then in a Branch-and-Bound algorithm. The space of search is drastically reduced and the Backtrack can immediately stop if a possible chain was found.

The algorithm starts with passing the requested TSD to the Intelligent Resource Selector. It tries to combine one matching MSAP to the Teleservice Descriptor, employing the same matching criteria as in the *single resource selection*. During the comparison of MSAPs only a subset of all attributes is taken into account. Figure 2 shows the corresponding attributes during a converter chain calculation. If a matching MSAP is found, the algorithm tries then to find an appropriate SAP matching its output.

In the negative case the algorithm continues trying to combine the next MSAP to the last found, otherwise the found SAP is the end of a possible converter chain. The found chain can be returned to the IReS calling entity.

If all converters have already been tested and no terminal was able to add to the converter chain the calculation is unsuccessful, so that no connection can be set up. Figure 7 shows a general overview of the developed algorithm.

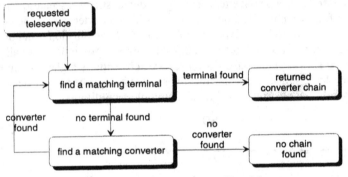

Fig. 7.   **Dynamic Resource Selection – Algorithm**

4.2.4. Calculating the most appropriate chain

With the algorithm introduced above it is possible to calculate a feasible converter chain. The algorithm stops if one solution was found. For calculating the most appropriate converter chain the same algorithmic approach could be used. Therefore we continue the calculation for finding all possible converter chains. The temporary solutions are stored and analyzed later. Due to its underlying Backtrack-algorithm the algorithm is able to find all possible converter chains without parsing the whole tree. Useless branches are detected early and will not be further analyzed.

The next step is now to calculate the QoS for every single temporary solution. The QoS of a Mediator Service Access Point has been introduced as a the QoS reduction through a converter. On this assumption, the algorithm can calculate the reduction of the QoS through a whole converter chain by combining the attributes of all the included components as an arithmetic procedure. An example is given in Figure 8.

A possible set of rules for calculating the overall QoS is given below, more details are discussed in [25].

- The overall value for "bandwidth" is the minimum "bandwidth" in the chain because an information flow is only as fast as at the slowest part on its way.
- The overall value for "delay" is the sum total of all delays in the chain.
- The overall value for "cost" is similar to "delay", adding up all components.

- The overall value for the attribute "intelligibility" results from the product of the percentage values, as

$$QoS_{overall} = \frac{QoS_1}{100} \cdot \frac{QoS_2}{100} \cdot \ldots \cdot \frac{QoS_n}{100} \cdot 100$$

After all $QoS_{overall}$ are calculated for all converter chains the algorithm searches for the best one, considering the user preferences as described previously.

A possibility for optimization the whole selection process is to evaluate the Quality of Services "on the fly". The resulting temporary $QoS_{overall}$ can be compared with the desired Teleservice Descriptor. If it falls below the requested one the calculation can be stopped, reducing the space of search more efficiently than the first one.

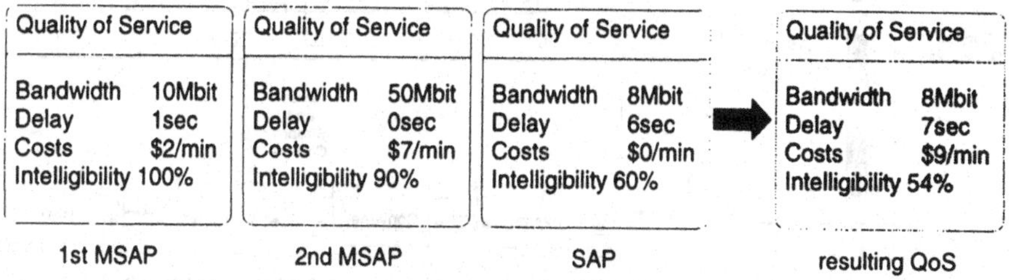

Quality of Service		Quality of Service		Quality of Service		Quality of Service	
Bandwidth	10Mbit	Bandwidth	50Mbit	Bandwidth	8Mbit	Bandwidth	8Mbit
Delay	1sec	Delay	0sec	Delay	6sec	Delay	7sec
Costs	$2/min	Costs	$7/min	Costs	$0/min	Costs	$9/min
Intelligibility	100%	Intelligibility	90%	Intelligibility	60%	Intelligibility	54%
1st MSAP		2nd MSAP		SAP		resulting QoS	

Fig. 8. **Dynamic Resource Selection – Quality of Service Combining**

If a valid converter chain was found, the included converters have to be configured. This means, that streams have to be connected, the QoS parameters have to be controlled and the connection has to be managed up to the end of the session. The control of customer premises equipment is not trivial. Within our project this task is handled by the Resource Configurator and the Converter Framework's Job and Stream Control, which are not covered by this paper.

5. Design and Implementation

In the first stages of this research project, the work in the iPCSS project was closely related to the complementary project of Personal Communication Support in TINA. During the ongoing work it was found that it would be more useful to focus on the aspects of media conversion within this project. Nevertheless, the reusability and possible integration into TINA was never lost from our perspective, and most design concepts of the current iPCSS consider this migration path from the beginning. The design is oriented on the Information Viewpoint and the Computational Viewpoint of the Reference Model for Open Distributed Processing (RM-ODP) [14]. In this paper only an abstract Computational Model (Figure 5) is given without all the performed decompositions of objects.

5.1. Computational Model

A Computational Model describes a system in terms of interacting Computational Objects (programming entities), see also the TINA Computational Modelling Concepts [9]. Computational objects (COs) have interfaces to communicate with other objects, namely operational interfaces and stream interfaces. A computational specification considers objects, the interfaces they support, and which interfaces they require at other objects.

The "Personal Communication related *user data*" CO is discussed in [24]. Recapitulating, the CO covers other COs, which contain information like user location, per-

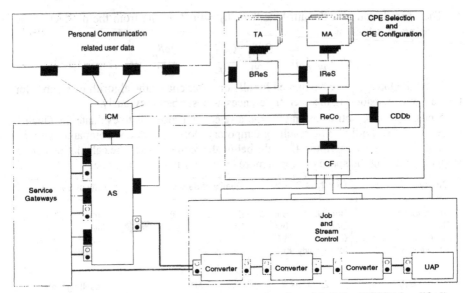

Fig. 9.   Resource Selection within the computational model of the iPCSS (simplified)

sonal schedule, call logic, and terminal registration. The Terminal Agent (TA) Computational Object models the SAP from section 4, representing a *terminal* of a user system. It maintains the capabilities of a specific terminal from the system perspective, to be considered in terminal selection activities done by the Basic Resource Selector. The description of *converters* is represented by the Mediator Agent (MA), modelling the MSAP from section 4. The Computational Object MA includes, conform to the TA, a Teleservice Descriptor for representing the supported service.

The Basic Resource Selector (BReS) is intended to maintain information about communication capabilities at certain locations. The BReS will maintain knowledge about a pool of resources, and therefore the BReS has to contain associations between locations (e. g. rooms or pico cells) and terminals, represented by the Terminal Agents (TA). The latter hold the terminal specific information, i. e. terminal state and capabilities referenced by the BReS.

A BReS dynamically selects a physical terminal depending on the requested service capabilities. Therefore it provides an intelligent selection algorithm to *find the most appropriate terminal*. Needed parameters for the selection are modelled in the TAs. The result of the BReS activities will be one terminal ID. This information will be returned to the intelligent Communication Manager (iCM). The Local Context only pertains to the user's current location to get terminal information appropriate to the requirements of a requested service.

The Intelligent Resource Selector (IReS) is an Service Supporting Object for the BReS. The Intelligent Resource Selector maintains knowledge about a pool of terminals/applications (TAs) and converters (MAs). Therefore the IReS has associations to Terminal (TA) and Mediator (MA). An IReS dynamically calculates the *most appropriate converter chain* depending on the requested Teleservice. The selection algorithm could be influenced by user preferences covered by the CO "Personal Communication related user data".

The Resource Configurator (ReCo) is an Service Supporting Object for the IReS. It has the ability to configure converter chains and to control the stream binding of them.

Therefore, an abstract MA/TA oriented connection graph (including the whole converter chain) is delivered to the ReCo, which has to be configured. The ReCo controls the configuration activities via the Converter Framework (CF). Only the CF has the knowledge and the ability to configure physical devices. The dedicated parametrization for devices is stored in the Converter Description Database (CDDb).

The intelligent Communication Manager (iCM) is the central component of the system. Triggered by the Active Store (which covers all the incoming gateways to the system) with a service request, it controls all the system behaviour. The functionality is defined like a state machine acting on dedicate system states.

In face of the distributed and object oriented approach the decision was made to use a CORBA conform platform for the implementation. The interfaces of the COs described in this paper are all specified with the Interface Definition Language (IDL). The used ORB is ORBIX on Sun Workstations (Solaris 2.5). The used C++ compiler is SUNWspro 4.2.

The runtime system comprises one entity of the Active Store, intelligent Communication Manager, intelligent Resource Selector, Resource Configurator, and Converter Framework for a service Request, respectively. The number of entities of Basic Resource Selectors, Terminal Agents, and Mediator Agents depends on the size of the represented communication scenario. Typical peculiarities including some offices with typical numbers of computers, telephones, and other communication devices, invoke several thousand entities of CORBA objects, as discussed in section 4.2.3.

The introduced systems is tested on computing resources and performance with the result, that a SUN Ultra 1 configured with 196 MB memory can handle several communication requests nearly real time. The only delay occurs during the starting process of any software, or the computing time of any asynchronous converter.

6. Summary and Outlook

This paper has presented the possibility of unified handling of telecommunication services by using the generic modelling of an Teleservice Descriptor, telecommunication devices, and service converters. Algorithms for finding appropriate terminals or converter chains for a communication request have been developed and implemented within the context of the iPCSS, representing a CORBA-based platform for the provision of full PCS capabilities, considering quality aspects for the automatic process.

The generic description, the algorithms developed, and the modular, scalable design of the iPCSS allows tailoring for various purposes and environments. While the testbed implementation is focused on the Customer Premises Equipment, the scalability in all directions, from global telecommunication architectures like TINA and IN (see below), down to a compact, individual solution, was a major aspect in the research [24]. The tool enabling such a wide range of scalability and independence from single/ multiple host environments was the usage of a middleware platform hiding such properties, CORBA in our case.

Intelligent Networks (IN), successful in telecommunication worldwide [19], know the concept of Intelligent Peripherals and Service Nodes (IP/SN) for the dialogue with the user and for enhanced service platforms. The iPCSS with its concept of the Active Store and the dynamic selection and configuration of individually tailored converter chains provides a powerful IP/SN [22, 23]. As it is already implemented on CORBA technology, the same migration strategies can apply as described for the TINA service nodes, as follows.

The Telecommunication Information Networking Architecture (TINA) represents a fundamental software architecture for Distributed Processing Environments (DPE), based on CORBA technologies, for rapid and flexible introduction of innovative services into information networking [7]. The CORBA based approach of the iPCSS and the design of the Computational Objects follow the ideas of the TINA Service Architecture [10] closely, with the purpose of allowing an integration of dedicated aspects of the iPCSS into TINA.

The emerging TINA Service Node concept, which provides an important step in evolution from IN based service environments toward TINA based service environments, follows the principal to model service control and service management TINA conform, but still provides access to the logic of IN based switches (via an IN Adaptation Unit) and TINA end-systems (via a TINA Access Session) [22].

Further information about our research can be obtained from http://www.fokus.gmd.de/ice/.

7. References

[1] ETSI Techn. specification TS 22.01v.3.1.0: Universal Mobile Telecommunication Systems (UMTS). Service Aspects, Service Principles. - Sophia Antapolis, France, 1997
[2] ETSI Draft 22.70v.0.0.3: Virtual Home Environments. - Sophia Antapolis, France, 1997
[3] IEEE Communications Magazine, Special Issue on Computer Telephony, Vol. 34 No. 4, April 1996
[4] Schmandt, Chris: Multimedia Nomadic Services on Today's Hardware. - in IEEE Network, Sept./Oct. 1994, pp. 12-21
[5] ITU-T Draft Recommendation F.851: Universal Personal Telecommunications - Service Principles and Operational Provision. November 1991
[6] Eckardt, T; Magedanz, T; Ulbricht, C; Popescu-Zeletin, R: Generic Personal Communications Support for Open Service Environments. IFIP World Conference on Mobile Communications, Canberra, Australia, September 1996
[7] Magedanz, T: TINA - Architectural Basis for Future Telecommunications Services. - in: Computer Communications Magazine, to appear in autumn 1997
[8] TINA-C Baseline. Service Architecture. Version 5.0, June 1997
[9] TINA-C: Computational Modeling Concepts. Vers. 3.2, TINA-C Stream Deliverable, TP_HC.012_3.2_96, Mai 17, 96
[10] Eckardt, T; Magedanz, T; Schulz, M; Stapf, M: Personal Communications Support within the TINA Service Architecture - A new TINA-C Auxiliary Project. - in: Proc. of 6th TINA Conference, Heidelberg, Germany, September, 1996
[11] BERKOM Project "Intelligent Personal Communication Support System", Deliverable 3: Resource Selection, CORBA Platforms, TINA migration. - German National Research Center for Information Technology (GMD), Research Institute for Open Communications System (FOKUS), Berlin, Aug. 1997
[12] Deutsche Telekom Project "Personal Communications Support in TINA", Report No. 1. - German National Research Center for Information Technology (GMD), Research Institute for Open Communications System (FOKUS), June 1996
[13] James Rumbaugh [et. al.]: Object-oriented Modeling and Design. Prentice Hall, Englewood Cliffs, New Jersey, 1991
[14] Raymond, K: Reference Model of Open Distributed Processing (RM-ODP): Introduction. - Proc. of the International Conference on Open Distributed Processing, ICODP'95, Brisbane, Australia, 20 - 24 February, 1995
[15] Mowbray, Thomas J.; Zahavi, Ron: The essential CORBA, System Integration using Distributed Objects, Jon Wiley & Sons, Inc., 1995, ISBN 0-471-10611-9
[16] Knuth, D.E.: The Art of Computer Programming. Vol.3: Sorting and Searching. - Addison-Wesley Publishing Co., Reading, Mass., 1973
[17] Ralf Steinmetz: Multimedia - Technologie: Einführung und Grundlagen. Springer-Verlag, Berlin, Heidelberg, 1993
[18] Murray, James: Encyclopaedia of Graphics File Formats. - O'Reilly: New York, 1994
[19] Magedanz, T.; Zeletin, R: Intelligent Networks. Basic Technology, Standards and Evolution. - London, etc.: Int. Thomson Computer Press, 1996
[20] ITU-T Recommendations Q.121x series: Intelligent Network Capability Set 1. Geneva, 1995
[21] ITU-T Recommend. M.3010: Principles of a Telecommunications Management Network. Geneva, 1992
[22] Rieken, R.; Carl, D.: Provision of integrated services - the Service Node approach. - Proc. of the IEEE Intelligent Network Workshop, IN'97, May 4-7, 1997, Colorado Springs, publ.: Los Alamitos (USA), IEEE Computer Society Press
[23] Leconte, A.: Emerging Intelligent Peripherals (IP) Applications for IN. - Proc. of the 3rd Int. Conference on Intelligence in Networks, ICIN'94, Oct. 11-13, 1994, Bordeaux
[24] Pfeifer, T.; Magedanz, T.; Popescu-Zeletin, R.: Intelligent Handling of Communication Media, Proc. of 5th IEEE Workshop on Future Trends of Distributed Computing Systems (FTDCS'97), Tunis, Tunisia, Oct 29-31, 1997
[25] Pfeifer, T.; Gadegast, F.; Magedanz, T.: Applying Quality-of-Service Parametrization for Medium-to-medium Conversion. - Proc. of the 8th IEEE Workshop on Local and Metropolitan Area Networks, Potsdam, Aug 25-28, 1996, publ.: Los Alamitos (USA), IEEE Computer Society Press
[26] Pfeifer, T.; Popescu-Zeletin, R.: Generic Conversion of Communication Media for supporting Personal Mobility. - Proc. of the 3rd COST 237 Workshop on Multimedia Telecommunications and Applications, Barcelona, Nov. 25-27, 1996, publ: Lecture Notes on Computer Science Vol. 1185, Springer: Berlin 1996
[27] A reference list of further relevant standards (ITU, CCITT), protocol descriptions, file formats, RFCs etc. is available on request.

AMnet: Active Multicasting Network

Ralph Wittmann and Martina Zitterbart

Institute of Operating Systems and Computer Networks
Technical University of Braunschweig
38106 Braunschweig, Germany
{wittmann|zit}@ibr.cs.tu-bs.de
http://www.ibr.cs.tu-bs.de/~wittmann/AMnet.html

Abstract. Today we are faced with an increasing variety of networks and end systems. The resulting heterogeneous environment imposes new challenges on communication support for multimedia and collaborative applications. AMnet is an approach that provides multipoint communication support for large-scale groups with heterogeneous receivers. Active network nodes are used to deliver data streams with user-tailored QoS. They provide so-called QoS filters that remove information from continuous media streams in order to reduce data rate for low-end receivers without affecting high-end receivers. A prototype implementation of AMnet based on RSVP is presented.

1 Introduction

In the past decade a tremendous growth in the use of computer networks in general and the Internet in particular for communication, computation, and collaboration can be noticed. Especially, an increasing demand for multimedia and collaborative applications, like teleworking and teleconferencing systems can be observed. For many of these applications group communication is a natural paradigm. Therefore, proper support within the communication subsystem is required. This includes flexible and scalable multicast support, enhanced group management, and provision of integrated services by the network.

Multicast support is inherently needed in order to facilitate development and to increase communication efficiency. This demand is acknowledged by various recent approaches that address transport protocols for reliable multicast. The focus of most current proposals is on the support of reliability. Mostly, all group members experience the same level of service, independent of their network attachment and end system equipment.

However, heterogeneity is growing with respect to networks and end nodes. Consider a scenario where some receivers of the group use simple PDAs, which typically are poorly equipped and often connected over wireless links (cf., Fig. 1). Others may use high speed workstations. Due to limited processing power low-end receivers might be unable to handle the same data stream as receivers using high-end workstations. Moreover, these devices may have small displays offering low resolution and color depth. Furthermore, group members do not only differ in terms of end-system capabilities, they may also be connected to networks with different characteristics. For example, an

ISDN network offers much less bandwidth than an ATM network. Hence, different participants in a multi-point communication may have different service requirements with respect to quality of service (QoS). The goal of the proposed research work is in the user-tailored service support for individual group members. Different data streams will be provided, e.g., to the high end workstation and to the wireless attached end node. In most current approaches the service provided to individual group members is penetrated by the group member with the lowest service capabilities. Such an approach is not acceptable for multimedia and collaborative applications in heterogeneous networking environments.

AMnet especially addresses the aspect of heterogenous group communication. It is based on Active Networking [11] in the sense that it uses active network nodes in order to individually tailor data streams to the end users service requirements. Therefore, so-called QoS-filters are implanted in network nodes. They adapt continuous media streams (e.g., MPEG-video) to specific QoS requirements.

The paper is structured as follows. Section 2 presents the AMnet approach and gives an overview of related work. A prototype implementation is presented in Section 3. It is based on RSVP as signalling protocol for QoS filter functions. Section 4 concludes the paper and gives an outlook on future work.

2   Heterogeneous Multicasting

A key aspect of AMnet can be seen in the usage of active networking in order to provide user-tailored services for heterogeneous groups.

2.1   The AMnet Approach

In Fig. 1 a multicast tree is depicted that provides heterogeneous multicasting capabilities through active network nodes, called *Active Multicasting Nodes*. Some of the network nodes included in the dissemination tree are active, others are passive with respect to enhanced multicasting tasks. Data streams with different QoS are delivered to the end users of the group, with QoS A > QoS B > QoS C. Therefore, the active network nodes comprise service modules for QoS filtering and enhanced QoS signalling. Furthermore, error control capabilities are implanted in the active network nodes.

The service modules can be selected and configured according to requirements of individual group members. Besides filtering and error control, ordering schemes, such as partial and causal ordered delivery, are other typical examples for selectable modules. Throughout this paper, we focus on QoS filters and the corresponding filter signalling.

In order to configure and activate the service modules inside the active network nodes, a signalling protocol is needed. With respect to signalling of QoS filters, the following requirements have to be met by the signalling protocol. A QoS filter should be placed as close as possible to the source of the multicast tree. All nodes above the selected node should have experienced higher QoS filter requests. Furthermore, dynamic join and leave operations must be supported. Therefore, dynamic allocation and re-configuration of QoS filters needs to be supported.

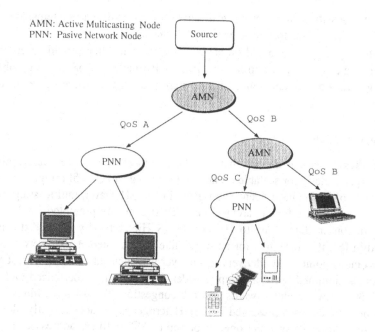

AMN: Active Multicasting Node
PNN: Pasive Network Node

Fig. 1. Multicast Tree with Active Multicasting Nodes

The active network nodes are also capable of supporting error control within the network. Therefore, multicast receivers are subdivided into smaller groups each served by an active network node that performs error control functions. A negative acknowledgement scheme is used. The active network node collects the NACKs, processes them and, if necessary, passes a NACK to an active network node that is located closer to the sender of the multicast tree. The sender only receives NACKs from active network nodes of the highest hierarchy level.

Active network nodes can also perform error recovery. If – in case of a NACK – the requested data are still cached inside the respective network node, it retransmits the data to the receiver. If not, the NACK is passed to active network nodes at higher hierarchy levels. This releaves in many cases the burden of retransmissions from the source.

Typically, retransmissions are not used in video conferencing applications. However, they can still be advantageous for such applications. Due to severe time constrains retransmission of continuous media from the source to the receiver is often impossible, because of the latency incurred. Since active network nodes can provide retransmissions, the experienced retransmission delay can be significantly reduced compared to the end-to-end delay.

Filtering of continuous data streams is closely related to synchronization. Many multimedia and collaborative applications need support for inter-stream synchronization. For example, a video-conferencing tool which synchronizes audio and video streams of a speaker. Support through the transport system is particularly needed if QoS filtering is used. Like error control, synchronization support has to be aware of QoS filters. This preserves the possibility to synchronize different streams in the end-systems. Moreover, the same argument holds for inter-stream synchronization.

In summary, AMnet is based on active network nodes, that provide user-tailored services for heterogeneous group communication. Particularly, QoS filtering, enhanced signalling, error control, and synchronization are provided by active nodes inside the network.

2.2 Related Work

AMnet addresses various aspects that have been subject of current research projects. This holds particularly for scalable reliable multicasting and QoS filtering.

Several QoS filters have been investigated in [15]. However, multicasting has not been addressed. Filters for hierarchical encoded streams are presented in [14]. They adapt continuous media streams to a specified QoS. Hierarchically encoded streams are also used in [8]. It is designed for networks like todays Internet, i.e., networks with best effort multi-point packet delivery. Each layer is associated with a multicast group. Receivers can join and leave a group in order to change the experienced QoS. This approach is mainly targeted towards network congestion. The usage of filters to control network congestion is presented in [13]. Heterogeneity is not explicitly addressed. Application level gateways, that operate on top of RTP [10] are addressed in [1] for filtering conversion operations. Reliable multicast is not addressed in this work.

The problem of providing multicast communication to a large group of receivers is a vital research area in the networking community. Especially, approaches, such as LBRM [5] and LGC [4] are similar to the concept followed by AMnet. However, these protocols do not support heterogeneous receivers. Moreover, QoS filtering of data streams may violate data integrity. Therefore, error control and recovery has to be aware of QoS filters.

An overview of the state-of-the-art in active networking can be found in [12]. Bhattacharjee et. al. [2] use active networking technologies for congestion avoidance. Aspects, such as multicasting or advanced QoS filtering are not addressed.

3   The RSVP-based AMnet Prototype

A prototype of AMnet has been implemented. In the context of the Internet, signalling protocols are currently investigated with the goal of providing an integrated services architecture [3]. RSVP [17] seems to be the main candidate for deployment in the Internet. Since its design is flexible with respect to future extensions, the prototype AMnet implementation is based on RSVP.

3.1   QoS filter

With AMnet QoS filter are used to adapt media streams to individual QoS requirements of heterogenous receivers. The focus was on compressed video streams that are MPEG-1 encoded. The implementation and integration of an H.261 video is currently under way. Other filters can be found in [16]. The following types of MPEG-1 filter are implemented [7]:

- Frame dropping filter
- Re-quantization filter
- Monochrome filter
- Slicing filter

The *frame dropping* filter simply drops P and B frames of the MPEG-1 stream to reduce the frame rate for less powerful receivers.

The *re-quantization filter* operates on DCT-coefficients and, thus, requires semi-decompression of the video stream. With re-quantization many near-zero coefficients may become zero which leads to a better compression ratio in the subsequent entropy encoding step. With moderate quantization steps a good trade-off between bandwidth reduction and loss of play-out quality can be achieved. However, large quantizers can result in strange artefacts [9].

The *Monochrome filter* removes color information. As MPEG encodes chrominance and luminance information independently, discarding of luminance components is a relatively simple operation.

Slicing filters exploit a property of MPEG encoded streams. The macro blocks of MPEG can be structured into *slices*. A slice contains information to synchronize the codec. If a slice gets lost, the codec resynchronizes on the following slice and may continue decoding. Since slices are not always included in MPEG coded data, the filter have to incorporate them. This may lead to a slightly higher data volume. However, slicing filter can drop slices easily and, thus, reduce the bandwidth consumed on the output link.

These types of QoS filters are configured in the active network nodes. In addition, a set of parameters is defined which specifies the requested filter function. The parameters are *quality factor*, *skip-frames*, *B/W-mode* and *number of slices*. The quality factor corresponds to a quantization level. The filter supports the quality factors 1 to 5, with 1 representing the best play-out quality, i.e., the least bandwidth reduction. The parameter skip-frames determines the frames to be dropped. The parameter B/W-mode activates the color reduction. Finally, the slice parameter specifies the number of slices per frame that the filter has to insert into the MPEG stream.

3.2 Signalling QoS Filter

Since QoS filters are located in the network, signalling is needed for configuration and allocation of QoS filters and for negotiation of filter parameters. As QoS filter reside on intermediate systems, they can be viewed as network resources. Due to its flexible concept, RSVP can be extended to configure and control QoS filters. In the current AMnet prototype RSVP serves as signalling protocol.

RSVP provides the concept of classes that can be defined in order to extend RSVP to new resources. The format of the new RSVP class *QoS filter* (QF) is depicted in Fig. 2(a). To assure compatibility with RSVP entities which are not extended to handle QF objects, the QF class carries the class number 208. With this class number a QF object is simply forwarded by an RSVP entity if the QF class is unknown. That way not every node must have a QoS filter Daemon running. To select a certain type of filter class the type field is used. As an example, Fig. 2(b) depicts a filteragent object for the implemented MPEG filter (cf., above).

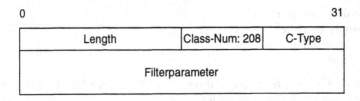

(a) Generic QF Class

Length: 12	Class-Num: 208	C-Type: MPEG
Picture-Quality: 3	Skip Frames: 0	
Slices per Frame: 2	Color/B\W-Mode: 0	

(b) Example: MPEF Filter Class

Fig. 2. RSVP Filter Class

3.3 Basic implementation architecture

The basic implementation architecture is depicted in Fig. 3. It consists of the following main components:

– RSVP Daemon
– QF Daemon
– QoS Filter (QF)

The RSVP daemon has been extended with an interface to the QF daemon to exchange QF objects [6]. In case a QF object is included in the received RESV message, the QF object is extracted and forwarded to the QF daemon.

The QF daemon is responsible for allocation and configuration of the QoS filter according to the requirements stated in the QF object. If it receives a QF object from the RSVP daemon, it needs to check whether other QoS requirements for that group have been received previously. Therefore, the QF daemon keeps some state information concerning filter requirements of group members. The following cases can be distinguished:

– no previous request
– previous request for same QoS filter type
– previous request for other QoS filter type

If no previous request is available, the RESV message is forwarded to the next node in the multicast tree.

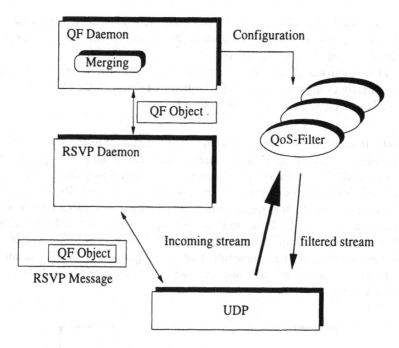

Fig. 3. Filter Agent Architecture

If another filtering request for the same group and the same QoS filter is available, merging of the filter parameters is needed. The QF daemon updates the QF object accordingly and forwards it to the RSVP daemon for further propagation towards the sender using RESV messages and standard RSVP procedures.

If a previous request for a different QoS filter type is registered at the QF daemon, two cases need to be distinguished.

First, the new request can be merged with the current request and the resulting request is not higher than the existing request. In this case, the corresponding filter functionality is activated by the QF daemon. The QF object that is periodically retransmitted with RSVP messages needs not to be updated. The node requires still the same amount of input data from the hierarchically higher nodes in the multicast tree.

Secondly, the new requested QoS filter can not be merged with the active filter. In this case, different filters serve different group members in the same active network node. To receive a sufficient data stream the highest requirements for both requests have to be identified. They form the new QF object to be forwarded in the RSVP messages.

A joining group member with new requirements can also lead to the re-location of already installed QoS filters. If the data is already filtered to a lower rate as the requirements of the joining group member, the filter may need to be moved to an active network node that is closer to the group members.

RSVP takes a soft state approach to managing the reservation state in routes and hosts. A soft state is created and periodically refreshed by PATH and RESV messages. QF objects are included in these messages.

QoS filters are released if group members leave the group and filtering at that node is no longer required. The same timeout mechanism used by RSVP is applied for that purpose. Thus, AMnet currently only provides soft-states.

RSVP daemon and QF daemon are involved in the control flow only. User data are directed directly to the corresponding QoS filter. The output of the QoS filter is then forwarded to the packet scheduler of the output link.

Example In order to illustrate configuration and merging of filters, the example depicted in Fig. 4 is considered. There are three receivers: A, B and F. They participate in a multicasted MPEG-1 video stream. Receiver A has severe bandwidth constraints due to a wireless link and, thus, requests a very low quality factor of 5. Receiver A issues a RESV message that includes the corresponding QF object. No other RESV messages have passed nodes C and D. Thus, the RESV message is forwarded to Node E. Node E has already seen a RESV message without bandwidth constraints from receiver F. As a result, node E has to install a new filter that is configured to the quality factor of 5 (cf., Fig. 4(a)), i.e., receiver F continues to receive the full stream and node D receives a reduced data rate.

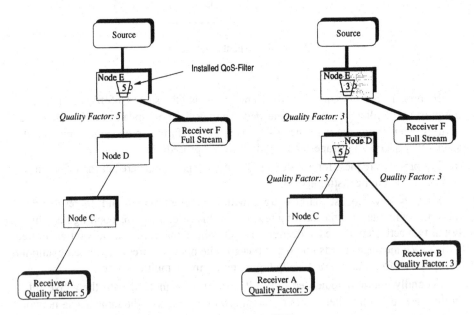

Fig. 4. RSVP and QoS Filter

In a subsequent step, receiver B joins the session. B requests a video stream with a moderate quality factor of 3 (cf., Fig. 4(a)). The RESV message of receiver B is forwarded to node D. Node D needs to merge the different filter requests from receivers A and B. As a result, a RESV message with a requested video quality factor of 3 is forwarded to node E. Moreover, a new filter is activated at Node D, which performs filtering according to quality factor 5. Due to the new request at node F, the filter is re-configured by the QF daemon to quality factor 3.

Such a simple merging style is possible, since the discussed filter types are not additive, i.e., applying the filter twice does not change the result.

Test scenario Fig. 5 depicts a test scenario for the prototype implementation. It shows two receivers with different QoS requirements. A test application permits to create a RSVP session and to request the QoS filter operations discussed above.

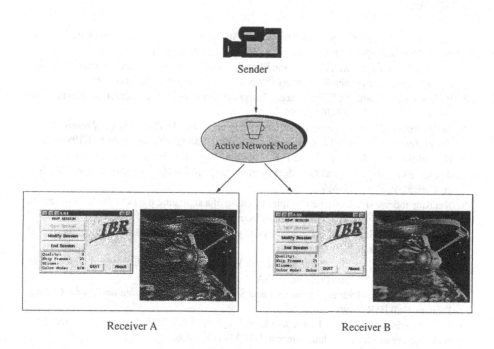

Fig. 5. Test Environment

AMnet is implemented on Sun Sparc Stations running SunOS 4.1.3 and Sun Ultra-Sparc Stations running Solaris 2.5 in our 100Mbit local Ethernet. In a first evaluation of our prototype implementation a MPEG-1 movie with a resolution of 120*160 pixels can be transmitted, filtered, and displayed with an average frame rate 12 frames per second. Without AMnet 15 frames per second are attained. A detailed analysis of the performance of AMnet is underway.

4 Conclusion

The paper addresses the issue of multicasting communication support for heterogeneous group communication. Typical applications that require such services are multimedia and collaborative applications. AMnet is an approach that is based on active networking, i.e., network nodes provide specific services to individual users. Active network nodes implement QoS filters, signalling for QoS filters, error control and synchronization. The existing prototype basically considers QoS filters and RSVP-based signalling of QoS filters. Several types of MPEG-1 filters have been implemented and tested.

Future work will focus on the impact of QoS filtering on synchronization and error control.

References

1. Elan Amir, Steve McCanne, and Hui Zhang. An application level video gateway. In *Proc. ACM Multimedia '95*, San Francisco, CA, November 1995. ACM.

2. Samrat Bhattacharjee, Ken Calvert, and Ellen W. Zegura. An architecture for active networking. In *High Performance Networking (HPN'97)*, White Plains, NY, April 1997.

3. R. Braden, D. Clark, and S. Shenker. Integrated services in the internet architecture: an overview. RFC 1633, ISI, June 1994.

4. M. Hofmann. A generic concept for large scale multicast. In *Proceedings of International Zurich Seminar on Digital Communication (IZS'96)*. Springer Verlag, February 1996.

5. H.W. Holbrook, S.K. Singhal, and D.R. Cheriton. Log-based receiver-reliable multicast for distributed interactive simulation. In *Proceedings of SIGCOMM '95*, Cambridge, MA, August 1995. ACM SIGCOMM.

6. Kai Krasnodembski. Design and implementation of a signalling protocol for filter functions (in german). Diploma thesis, TU Braunschweig, August 1997.

7. Thomas Kupka. Scaling of MPEG-I encoded videostreams (in german). Diploma thesis, TU Braunschweig, August 1997.

8. Steven McCanne, Van Jacobson, and Martin Vetterli. Receiver-driven layered multicast. In *ACM SIGCOMM'96*, San Francisco, USA, August 1996.

9. K.R. Rao and J.J. Hwang. *Techniques and Standards for Image Video and Audio Coding*. Prentice Hall PTR, 1996.

10. H. Schulzrinne, S. Casner, R. Frederick, and V. Jacobson. RTP: A transport protocol for real-time applications. Technical report, GMD Fokus, 1996.

11. D. L. Tennenhouse and D. J. Wetherall. Towards an active network architecture. *Computer Communication Review*, 26(2), April 1996.

12. David L. Tennenhouse, Jonathan M. Smith, W. David Sincoskie, David J. Wetherall, and Gary J. Minden. A survey of active network research. *IEEE Communications Magazine*, 35(1):80–86, January 1997.

13. Hartmut Wittig, Jörg Winckler, and Jochen Sandvoss. Network layer scaling: Congestion control in multimedia communication with heterogenous networks and receivers. In *International COST 237 Workshop on Multimedia Transport and Teleservices*, Vienna, Austria, November 1994.

14. Lars C. Wolf, Ralf Guido Herrtwich, and Luca Delgrossi. Filtering Multimedia Data in Reservation-Based Internetworks. Technical Report 43.9608, IBM European Networking Center, 69115 Heidelberg, Germany, August 1994.

15. N. Yeadon. *Quality of Service Filtring for Multimedia Communications*. Phd thesis, Lancaster University, May 1996.

16. N. Yeadon, F. Garcia, D. Shepherd, and D. Hutchinson. Continuous media filters for heterogeneous internetworking. In *Proceedings of the Conference in Multimedia Computing and Networking 1996*, San Jose, California, January 1996.

17. Lixia Zhang, Stephen Deering, and Deborah Estrin. RSVP: A new reavailable ReSerVation Protocol. novel design features lead to an Internet protocol that is flexible and scalable. *IEEE network*, 7(5), September 1993.

Impact of Virtual Group Structure on Multicast Performance

Markus Hofmann, Manfred Rohrmüller

Institute of Telematics, University of Karlsruhe, 76128 Karlsruhe, Germany
Phone: +49 721 6086413, Fax: +49 721 388097
{hofmann, rohrmuel}@telematik.informatik.uni-karlsruhe.de

Abstract. Scalability will be a key issue in the design and the development of reliable multicast protocols for the Internet. As the geographic span and the size of communication groups increase, efficient connection management schemes including scalable error and congestion control become more and more important. Besides other approaches, several schemes based on subgrouping have been proposed to overcome the well-known implosion problem and to optimize network utilization. However, the performance of these approaches strongly depends on the virtual group structure used for local error recovery and congestion control. While a certain structure may reduce average transfer delay, another one may be suitable to decrease overall network load. This paper discusses the suitability of several metrics for subgrouping of global multicast groups and investigates the impact of virtual group structures on the overall performance of multicast communication.

1 Introduction

Groups present an ubiquitous form of relationship and interaction in human society. People get together in groups to share common interests or to work on collaborative projects. Distributed computer systems are arranged into cooperative groups to master complex problems. Emerging applications, such as collaborative distributed work or information dissemination, rely on group interaction and are expected to require information exchange between a large number of geographically dispersed components. The recent success of applications deployed over the MBone [6] illustrates the enormous potential of group communication and demonstrates the instant need for economic multicast services in the Internet. Recently, multicast data transmission based on Deering's IP multicast extensions [1] has been widely available in the Internet. However, the bearer service provided by IP does not fit the requirements of each individual application. It offers a best-effort service leaving it up to the application to provide the required quality of service. Several error correction schemes have been proposed to improve reliability of multicast communication in the Internet [2]. All of them have to deal with the well-known implosion problem due to feedback messages generated by the receivers.

The *Scalable Reliable Multicast (SRM)* [3], for example, uses damping and slotting mechanisms to reduce state management overhead. Receivers solely take the responsibility for error correction, which is why SRM achieves a high degree of fault tolerance. SRM is an example of the receiver-based approach for error control. A receiver missing a certain data unit multicasts a repair request to the whole group. Group members that have successfully received the requested packet will multicast it to the entire group. To avoid a flood of repair requests and of retransmission, SRM suppress redundant requests by using timers carefully set and adjusted to the current network load.

Other approaches arrange receivers in a virtual tree hierarchy with the sender at the root. The *Reliable Multicast Transport Protocol (RMTP)* [7] or the *Tree-based Multicast Transport Protocol (TMTP)* [9], for example, represent a multi-level hierarchical approach in which leaf receivers periodically send status messages to the controller of their subgroup. The controllers, in turn, send their status periodically to the higher layer controllers. This scheme continues until the controllers at the highest level send their status directly to the sender. Lost data packets are always requested from a higher level controller, not making use of local retransmissions between neighboring receivers.

The *Local Group based Multicast Protocol (LGMP)*[1] [4] defines a hybrid approach. It supports reliable and semi-reliable transfer of both continuous media and data files. LGMP is based on the principle of subgrouping for local error recovery and local acknowledgment processing. Receivers dynamically organize themselves into subgroups, which are called Local Groups. They dynamically select a Group Controller to coordinate local retransmissions and to handle status reports. The selection of appropriate receivers as Group Controllers is based on the current state of the network and of the receivers themselves. However, the selection of Group Controllers is not a task of a data transfer protocol like LGMP. Instead, we have defined and implemented a separate configuration protocol, which we call *Dynamic Configuration Protocol (DCP)* [5]. Packet errors are firstly recovered inside Local Groups using a receiver-initiated approach. Missing data units are requested from the sender or a higher level Group Controller only if not even a single member of the Local Group holds a copy of the missing data unit. Otherwise, errors will be recovered by local retransmissions. Full reliability and efficient buffer utilization are ensured by a novel, three-state acknowledgment scheme.

All these approaches are based on the principle of subgrouping. Receivers are divided into separate subgroups, each of them represented by a Group Controller. The subgroups are arranged into a multi-level hierarchy defining a so-called *Virtual Group Structure*. The placement of controllers and the arrangement of subgroups will strongly influence the efficiency of multicast data transfer. While a certain Virtual Group Structure may be chosen to reduce network load, another one may increase average throughput.

The paper discusses the impact of group structure on the efficiency of subgroup-based multicast communication. It presents simulation results and illus-

---

[1] LGMP is based on the Local Group Concept (LGC)

trates in which way receivers should be arranged to optimize average transfer delay and overall network load. Section 2 introduces the MBone scenario all the simulation models are based on. The following subsections investigate the impact of subgroup size and group hierarchy on network load and average transfer delay. Finally, Section 3 presents a protocol for automated establishment of Virtual Group Structures and Section 4 concludes the paper.

2 Performance Evaluation

The multicast algorithms of the Local Group Based Multicast Protocol (LGMP) have been investigated by performing a large number of different simulations. However, the main conclusions are also valid for other subgroup-based or tree-based multicast approaches. Early results have provided feedback to the development and implementation of LGMP. All simulations have been performed using BONeS/Designer, an event-driven network simulation tool by the Alta Group of Cadence Design Systems. Each simulation model comprises a network topology, a protocol state machine and packet formats used for data and control message exchange. The multicast algorithms of LGMP have been compared to a common, sender-based multicast approach. In this approach, receivers address their acknowledgments directly to the multicast transmitter, and necessary retransmissions are performed solely by the sender.

In a first step, several simple scenarios have been modeled to get an idea about the scalability of LGMP. Multicast groups with up to 2000 receivers have been simulated. The simulation results show significant improvements compared to common multicast techniques [4]. Besides the investigation of scaling issues, the effect of different group structures and the placement of Group Controllers is of interest for further development of scalable multicast protocols. Therefore, more complex simulation models based on a real MBone scenario have been developed. The results of these comprehensive simulations allow us to draw conclusions on the suitability of different virtual group hierarchies. A recommendation on how to subgroup the members of a multicast group is given. These recommendations are also valid for tree-based multicast protocols such as RMTP or TMTP.

2.1 The Simulation Scenario

A major claim of the project is to get authentic and realistic statements on the impact of group structure on overall multicast performance. Instead of analyzing simple and abstract network scenarios, all our simulation models are based on packet loss data collected in the MBone by Yajnik, Kurose and Towsley [8]. The MBone scenario described in their paper and all the protocol mechanisms of LGMP are modeled with full details. This results in quite complex simulation models, each of them running more than two days on a SUN SparcStation 20 with two processors.

The scenario used in our simulations consists of a packet source located in California and of 11 receivers $r_{0,0}$ through $r_{10,0}$ distributed all over the world.

The hosts are connected via the MBone. The average packet loss rate of each link, as reported in [8], is inscribed in Figure 1. The bold lines represent the *backbone* links of the network. These links form the base of the multicast tree and traverse most of the distance in the network. All the other branches are on the *edge* of the multicast routing tree and connect receivers to the backbone. However, the edge branches may cross multiple multicast routers before reaching the receivers.

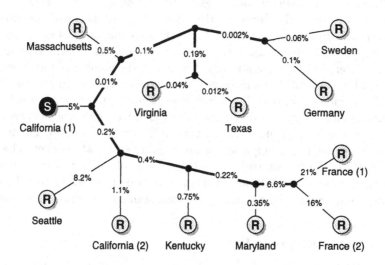

Fig. 1. MBone scenario used for simulations (based on [8])

In order to increase the overall group size, we assume 19 additional group members $r_{i,1}$ through $r_{i,19}$ ($0 \leq i \leq 10$) in the neighborhood of each (real) receiver $r_{i,0}$. Therefore, the number of simulated receivers is 220. However, the paper always refers to a certain receiving site as the complete set of all 20 simulated receivers. While talking about the receiver in Kentucky, for example, we refer to all the 20 simulated receivers within the area of Kentucky. The additional receivers $r_{i,1}$ through $r_{i,19}$ are modeled using the error probability and the transfer delay given by the receiver $r_{i,0}$. This assumption is quite realistic because the receivers $r_{i,1}$ through $r_{i,19}$ are assumed to be in the same local region and to be connected to the same branch of the multicast transmission path as receiver $r_{i,0}$. The expansion of group size does not effect subgrouping of receivers, because each of the 20 receivers will be assigned to the same Local Group.

All the other parameters are set according to the results presented in [8]. The transmission delay, which is not given in the paper, is set to 40 msec for packets transmitted within the USA and to 100 msec for packets transmitted between hosts in the USA and receivers in Europe. Transmission errors are simulated by Markov models using the average packet loss rate and the average burst length for input. The latter one is set to 2 packets which is in conformity with the results presented in [8].

The packet source transmits 1000 data units per second with a constant size of 2000 Byte per packet. This results in a data rate of about 2 MByte/sec. The simulation time is set to 40 seconds for each simulation. Hence, the sender emits 40000 data packets during the whole simulation. Note that this amount includes all the retransmissions performed by the sender. A status request is sent after each hundredth data packet.

Our simulations examine the average transfer delay and the overall network load. Normally, the *network load* is given by the number of packets traveling across the backbone links. However, the total number of data packets transmitted in our simulations is always set to 40000, as explained before. Therefore, the network load is given relatively to the number of packets originally sent by the transmitter. This relation strongly depends on the number of retransmission performed by the sender. An increase in the number of retransmissions always results in an increase of overall network load. Therefore, the relative value is a good measure for network load. *Transfer delay* is the average time between sending a data packet and its successful delivery to the receiver process. The transfer delay is determined by using a time stamp within each data packet. The time stamp is set by the transmitter on sending a packet for the first time. Because system time is globally synchronized in our simulation models, the transfer delay can easily be calculated by evaluating the time stamp of incoming data packets.

2.2   Impact of Subgroup Size

In order to investigate the impact of group size on multicast performance, subgroups of different size have been defined and evaluated. The global multicast group has been split into separate Local Groups, whereby different subdivisions vary in the maximum distance n allowed between two members of a subgroup. The metrics *Hop Count* and *Loss Probability* have been used to define this distance. The algorithm used to subdivide the receivers and to arrange them into subgroups is the following:

1. Define the maximum distance n between all the members of a subgroup and determine all the possible subdivisions.
2. For a given maximum distance, the solution resulting in a minimum number of subgroups is chosen.
3. If there are multiple subdivisions resulting in the same minimum number of subgroups, choose a solution with homogeneous subgroup sizes.

An example for each of these rules is given to illustrate the algorithm:

1. The receivers in *Massachusetts*, *Virginia*, *Texas*, *Sweden* and *Germany* are combined in a single subgroup if the maximum distance within a subgroup is defined to be four hops. A maximum distance of three hops results in three separate subgroups, namely {*Massachusetts*}, {*Virginia*, *Texas*} and {*Sweden*, *Germany*}.

2. There are two possible subdivisions of the receivers in *Seattle, California (2)*, *Kentucky, Maryland, France (1)* and *France (2)* if a maximum distance of three hops is given:
   - {*Seattle, California (2), Kentucky*} and {*Maryland, France (1), France (2)*}
   - {*Seattle, California (2)*}, {*Kentucky, Maryland*} and {*France (1), France (2)*}

   According to the second rule, the first subdivision resulting in two subgroups is chosen.
3. If the maximum distance is set to four hops, the receivers in *Seattle, California (2)*, *Kentucky, Maryland, France (1)* and *France (2)* could be divided into a subgroup of size four and a subgroup of size two. Another solution is to split them into two subgroups of size three. According to the third rule, the latter solution will be chosen.

Table 1 lists the resulting subdivisions according to these rules. The name of each subdivision indicates the metric that has been used to build up the subgroups: Subdivisions including a roman I have been built using the hop count metric. Subdivisions including a roman II are based on the loss probability. An f at the end of the name indicates a flat virtual group structure, whereby all the subgroups are directly attached to the sender. Later on, an h indicates a multi-level hierarchy. The paper makes intensive use of these naming conventions in order to explain the simulation results. Table 1 also includes the maximum allowed distance n between two members of a subgroup of each subdivision.

The impact of group size on multicast performance has been investigated using flat virtual group structures. All subgroups were directly connected to the sender. Multi-level hierarchies are discussed in the following section. In addition, a standard multicast simulation has been performed to illustrate the benefits of LGMP. In the standard multicast simulation all 220 receivers address their status reports directly to the sender. Furthermore, they request missing data packets always from the sender.

Figure 2 presents the overall number of data packets sent by the transmitter in relation to the number of original data packets. This value allows conclusions concerning the network load caused by reliable multicast transport. The average transfer delay observed by the receivers is given in Figure 3. It illustrates the delay of packets to all receivers within each subgroup as well as the average delay of packets all the members of the global multicast group.

The impact of group size on transfer delay and network load can be derived from the results of simulation *I.a-f* through *I.d-f*. As presented in Figure 2 and 3, the global network load as well as the average transfer delay decrease with an increase of average subgroup size. Especially receivers in *Sweden, Germany, France (1)* and *France (2)* observe a higher transfer delay in smaller subgroups.

Note that the existence of small subgroups results in a higher transfer delay than for the standard multicast communication. This can be seen in the simulation *I.d-f* and is due to the mechanism of local retransmission and local

Table 1. Simulated subdivisions to examine the impact of group size

Scenario	n	#LGs	Subdivision
I.a-f	5	2	$LG1$: { Massachusetts, Virginia, Texas, Sweden, Germany } $LG2$: { Seattle, California (2), Kentucky, Maryland, France (1), France (2) }
I.b-f	4	3	$LG1$: { Massachusetts, Virginia, Texas, Sweden, Germany } $LG2$: { Seattle, California (2), Kentucky } $LG3$: { Maryland, France (1), France (2) }
I.c-f	3	5	$LG1$: { Massachusetts } $LG2$: { Virginia, Texas } $LG3$: { Sweden, Germany } $LG4$: { Seattle, California (2), Kentucky } $LG5$: { Maryland, France (1), France (2) }
I.d-f	2	7	$LG1$: { Massachusetts } $LG2$: { Virginia, Texas } $LG3$: { Sweden, Germany } $LG4$: { Seattle, California (2) } $LG5$: { Kentucky } $LG6$: { Maryland } $LG7$: { France (1), France (2) }
II.a-f	1,5%	6	$LG1$: { Massachusetts, Virginia, Texas, Sweden, Germany } $LG2$: { Seattle } $LG3$: { California (2) } $LG4$: { Kentucky, Maryland } $LG5$: { France (1) } $LG6$: { France (2) }

acknowledgment processing. In large subgroups, the ability to recover from errors by local retransmissions is quite high. It is likely that one of the members has correctly received the missing data packet. Therefore, it is not necessary to request it from the sender. Especially receivers far away from the sender, such as *Sweden* and *Germany*, gain by local error recovery.

On the other hand, it is not likely that errors can be recovered locally in small subgroups. This is especially true if all the members of a subgroup are connected to a single subnet. In this case, packet errors are likely to be spatially correlated making it impossible to perform local error recovery. Instead, the controller of such a subgroup has to request missing data packets from the sender or its higher level group controller. This effect can be seen evaluating the average transfer delay of receivers *France(1)* and *France (2)* in simulation *II.a-f* (see Figure 3).

One may wonder that subgroup-based multicast communication can even be slower than standard multicasting. However, this drawback is due to the

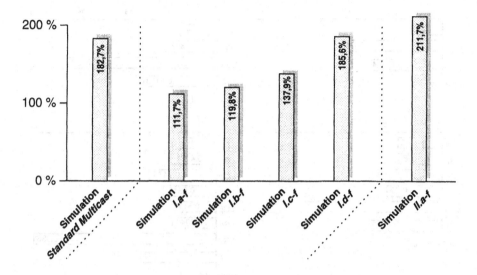

Fig. 2. Global network load for flat group structures

necessity to collect and process all the status reports of subgroup members. Therefore, retransmission requests directed to the sender are delayed until all the members of a subgroup have negatively acknowledged the missing data. To overcome this drawback, a new version of LGMP incorporates periodic status push instead of controller-initiated status poll [5].

The results show that small subgroups should be avoided in flat hierarchies. Multicast performance improves with an increasing number of subgroup members. However, there is always a trade-off between the size of a subgroup and the processing capacity required to handle it. Too many members will result in a controller implosion and will degrade multicast performance. The optimal size of a subgroup strongly depends on the processing capacity of its controller and of the error characteristics of its members. A controller should handle as many receivers as possible. Therefore, a mechanism is required to monitor the load of each controller and to dynamically adapt the virtual group structure to the current network load and receiver state. The Dynamic Configuration Protocol (DCP), presented in [5] and in Section 3, provides such a mechanism.

2.3 Examination of Different Group Hierarchies

To examine the benefits of multi-level hierarchies additional simulations have been performed. They have been based on the subdivisions presented in the former section. However, the subgroups have now been arranged in a multi-level hierarchy according to the given network topology (see Figure 1) and depending on the distance metric (Hop Count or Loss Probability). The resulting group hierarchies are given in Table 2.

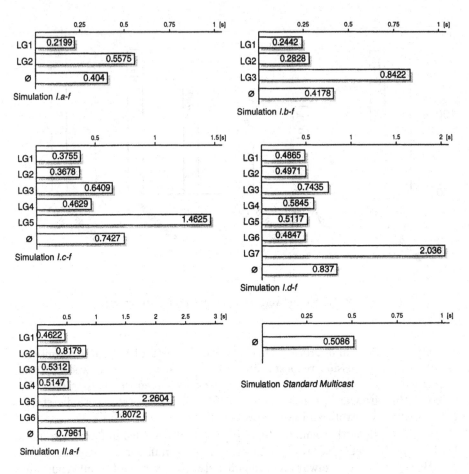

Fig. 3. Average transfer delay for flat group structures

Figure 4 presents the measured network load. As previously explained, it is given relatively to the number of packets originally sent by the transmitter. The average transfer delay is presented in Figure 5.

A comparison of simulation *I.c-f/h* and *II.a-f/h* demonstrates that multi-level hierarchies yield better performance than flat group structures. The evaluation of average transfer delays in group *LG2* and *LG3* shows better performance for the hierarchical group structure in *I.c-h* compared to the corresponding flat structure in *I.c-f*. This is also valid for the groups *LG5* and *LG6* in the simulations *II.a-f/h*. It is remarkable that the mentioned subgroups are exactly those placed in the second level of the hierarchy. These subgroups are not directly attached to the sender. Instead, they address their status reports to a higher level controller.

It is also interesting that in simulation *II.a-h* the receivers in Europe (members of *LG1*, *LG5* and *LG6*) observe nearly the same average transfer delay than receivers in the USA. This illustrates the benefits gained by multi-level hierarchies. Instead of addressing requests for retransmissions directly to the sender in

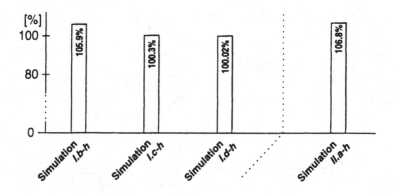

Fig. 4. Global network load for multi-level hierarchies

California, controllers in a multi-level hierarchy get missing data packets from a nearby, higher-level controller.

Figure 5 illustrates that the average transfer delays observed by the groups *LG1* and *LG4* in simulation *I.c-f/h* and by the groups *LG1*, *LG2*, *LG3* and *LG4* in simulation *II.a-f/h* decrease if the subgroups are arranged in a hierarchy. Furthermore, there is an assimilation of average transfer delay in those subgroups that are arranged in a multi-level hierarchy (such as the groups *LG1*, *LG2* and *LG3* in *I.c-h*).

A look at the global network load shown in Figure 4 reveals similar results as explained in the former section. The definition of multi-level hierarchies results in a decrease of network load. Retransmissions between controllers at different

Table 2. Group structures used to examine the impact of group hierarchy

Scenario	I.b-h	I.c-h	I.d-h	II.a-h
Parent of LG 1	California(1)	California(1)	California(1)	California(1)
Parent of LG 2	California(1)	LG1	LG1	LG3
Parent of LG 3	LG2	LG1	LG1	California(1)
Parent of LG 4		California(1)	California(1)	California(1)
Parent of LG 5		LG4	LG4	LG4
Parent of LG 6			LG4	LG4
Parent of LG 7			LG4	

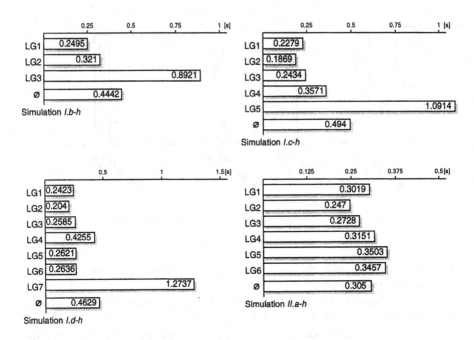

Fig. 5. Average transfer delay for multi-level hierarchies

levels reduce the number of data packets to be retransmitted from the far away sender in California.

The simulation results exhibit two essential conclusions on multi-level group structures:

- There is an assimilation of multicast performance in hierarchically arranged subgroups.
- The arrangement of a few subgroups in a hierarchy results in a better performance for all receivers.

These results are due to local error recovery and local acknowledgment processing. Only those packets that are lost by all the receivers of a subgroup will be requested from the sender. In a multi-level hierarchy, members of a subgroup could also be controllers of a lower-level subgroup. Therefore, the mechanism of local error recovery continues recursively. Only those packets missed by all the members of a subgroup and all the members of its children have to be retransmitted by the sender (or a higher level controller). This results in an assimilation of the performance and decreases the total number of packets to be retransmitted by the sender. Network load and average transfer delay are much less than in flat virtual group structures.

A closer look on the group $LG1$ in $I.b$-f and in $II.a$-f (which contain both the same set of receivers) exhibits another interesting fact. Subgroups with relative high packet loss rates (such as the groups $LG5$ and $LG6$ in $II.a$-f) make the performance in all the other groups worse. Performance is decreased due to the involved error control and congestion avoidance mechanisms. LGMP, as well as other multicast protocols, implements a rate-based congestion control with multiplicative decrease and additive increase of transfer rate. According to the defined algorithm, a sender will reduce its transfer rate due to status reports from subgroups with high packet loss rate. In addition, performance is decreased due to high load at the sender. While the sender performs retransmissions requested by subgroups suffering from high loss rates, all the other receivers are waiting for new data to arrive. This argument is confirmed by the network load measured for both simulations. Therefore, subgroups with a high loss probability should be attached to another subgroup rather than directly to the sender.

3   The Dynamic Configuration Protocol (DCP)

The results of our simulations provided feedback to the design and the implementation of a protocol named *Dynamic Configuration Protocol (DCP)* [5]. DCP provides mechanisms for an automated establishment of virtual group structures and for dynamic reconfiguration in accordance with the current network load and group membership. No manual administration is necessary. The definition of subgroups is based on a combination of multiple metrics depending on the QoS requirements of the user. DCP is self-organizing and tolerant with respect to failing controllers.

3.1   Expanded Ring Advertisement

Each Group Controller periodically sends packets of type LG_ADVERTISE to announce its existence. These messages are sent using a separate multicast address. Communication participants listen to the group-specific DCP address and use received advertise messages to identify existing Group Controllers. Advertise messages contain information that allows receivers to select the most appropriate Group Controller according to their requirements. By default, an advertise message includes the smoothed error probability of a Group Controller, the number of receivers currently controlled by the Group Controller as well as the multicast address of the represented Local Group. Optional fields have been defined to include additional information, for example a timestamp for the calculation of transfer delay between Group Controller and receiver.

There is some kind of trade-off between network load caused by advertise messages and the time required to react upon dynamic changes in group structure. Frequent sending of advertise messages ensures short reaction times while increasing the network load. Moreover, the visibility of Group Controllers is determined by the scope of their advertise messages. The larger the TTL value of outgoing advertise messages, the more receivers will be able to identify a Group Controller. On the other hand, advertise messages should preferably be limited to a local scope in order to avoid a global flood of control traffic.

To deal with these contrary requirements, we have designed a new mechanism called *Expanded Ring Advertisement*. Group Controllers send their advertise messages with dynamically changing TTL values according to Table 3. The first message is sent with a scope of 15 (scope 'Site'), the second one with a value of 31 (scope 'Region'), etc.

Table 3. TTL values used to send advertise messages

Interval No.	1	2	3	4	5	6	7	8	9	10	11	12	13	14	15	16	17	18	19
TTL	15	31	15	63	15	31	15	127	15	31	15	63	15	31	15	254	15	31	15

Receivers within a scope of 15 will get each of the advertise messages. If the distance of a host is between 16 and 31 hops, it will receive every second advertisement. This scheme continues in a way that every 16th advertise message will be distributed worldwide. The Expanded Ring Advertisement ensures that the frequency of advertise messages exponentially decreases with increasing scope. Therefore, the scheme reduces network load while allowing short reaction times upon changes within the local scope of a receiver. The Expanded Ring Advertisement could also be used to estimate the number of hops between a receiver and a Group Controller.

3.2 Selection and Placement of Group Controllers

Once the service user has issued a listen request, a receiver initializes an association control block. Each of these blocks contains an entry named redirect, which is undefined at startup. This entry will identify the controller to which receivers should deliver their status reports. While the value of redirect is undefined, LGMP will address all status reports to the data source.

With the establishment of an association control block, the receiver activates an initialization timer named INIT-TIMER and changes from the inactive to the pending state. It joins the group-specific DCP group and, therefore, stimulates the transmission of an IGMP Host Membership Report. Now, the host is set up to receive packets addressed to the global DCP group, and it buffers all the information obtained from received advertise messages. After expiration of timer INIT-TIMER, a receiving DCP instance evaluates the buffered information, selects one of the discovered Group Controllers, sets the redirect entry of the association control block to the address of the chosen Group Controller, and changes to the active state. While being in active state, a receiver will continue to process advertise messages and to update the redirect entry dynamically.

If no appropriate Group Controller could be found according to the application requirements, the joining receiver has two possibilities. On one hand, it could attach itself directly to the Local Group represented by the multicast transmitter. In this case, the redirect entry of its association control block will be undefined and LGMP will address all status reports to the multicast transmitter. On the other hand, it could establish a new Local Group and appoint itself as Group Controller of the new subgroup. One of the identified Group Controllers or the multicast transmitter itself will be defined to be the parent of the newly established subgroup. All reports about the status of the new Local Group will be addressed to the parent Group Controller, thus building a group hierarchy.

Initially, it is the founder of a Local Group which will become the Group Controller. Due to the joining and dropping out of receivers, the group structure has to be reconfigured dynamically during the lifetime of an association. It might be beneficial to split a growing Local Group or to merge several waning subgroups. In addition, a joining receiver might be a better Group Controller than the current one due to its network connection or its processing capacity. The following section describes the scheme used to perform such a dynamic reconfiguration of the global group structure.

3.3 Dynamic Reconfiguration of Local Groups

As receivers and Group Controllers may join and leave during the lifetime of a connection, it is necessary to adjust the placement of Group Controllers dynamically according to the current group status, the current network load, and the current characteristics of each communication participant. For example, it could be advantageous to place the Group Controller in the center of a Local Group. Various schemes based on different criteria could be used to determine the optimal Group Controller among all the members of a Local Group.

Besides the move of Group Controllers, the splitting and merging of Local Groups might become necessary due to changes in group membership. The burden of acknowledgment processing and of doing local retransmissions has to be distributed fairly among all the Group Controllers, thus, resulting in a well-balanced group structure.

Receivers use information contained in LG_ADVERTISE messages to maintain a table of reachable Group Controllers. On receiving an advertise message, a host will add a new entry to the table or update an existing one. Each entry represents a Group Controller and indicates its error probability, the estimated number of hops between receiver and Group Controller as well as size and multicast address of the Local Group. In case there is more information about the characteristic of a Group Controller included in its advertise messages, this information will also be added to the table (e.g. transfer delay or carrier fees). While updating their table, Group Controllers ignore their own advertise messages.

Each entry is valid for a time interval T_{val}. When the timer expires and no further advertise message of a certain Group Controller has been received within the last time interval, receivers will delete the corresponding entry in the table. Therefore, each host has an up-to-date view on active Group Controllers, their identity, and their current status. There is no additional information exchange necessary to keep the table valid. If a Group Controller fails or leaves, the corresponding table entry will time out and be deleted. To ensure correctness of this mechanism, the expiration time T_{val} must be longer than the time T_{adv} between two successive advertise messages of the Group Controller concerned.[2]

---

[2] The time T_{val} depends on the distance between a receiver and the Group Controller.

We propose to choose $T_{val} > (3 \cdot T_{adv}) + \epsilon$ to counterbalance the loss of two successive advertise messages. If three successive announce messages are lost, a receiver will erroneously delete the corresponding entry. However, on receiving the next advertise message the receiver will add the Group Controller again.

Receivers periodically rate the suitability of their current Group Controller GC_i. If the rating r_j of another Group Controller GC_j is better than the rating r_i, the redirect entry will be set to the address of GC_j. However, the difference between r_i and r_j should be higher than a given threshold to avoid oscillatory changing between Local Groups. A problem might also occur in case a large number of receivers decide to assign themselves to a newly defined Group Controller. Due to an overwhelming number of new members, the new Group Controller might get under heavy load, thus decreasing its rating. This would probably result in a further reconfiguration. Therefore, receivers delay spontaneous reconfiguration for a random time to check the rating of a newly detected Group Controller again.

In addition, a receiver R_i periodically calculates its own rating r_i. If r_i is better than the rating r_j of its current Group Controller by some non-negligible amount, the receiver will establish a new Local Group claiming itself to be a Group Controller. It will start to send LG_ADVERTISE messages to advertise its existence and its current status. Nearby receivers may now join the newly established Local Group, thus relieving their previous Group Controller.

4 Conclusions

Subgroup-based multicast protocols, such as RMTP, TMTP or LGMP, are designed to provide efficient and low-cost error control for multicast applications in wide-area networks. The simulation results presented in this paper give evidence that these schemes succeed in keeping retransmissions relatively local within the wide-area topology and in reducing sender implosion. These properties will allow scaling to large receiver sets in large-scale networks. However, the performance of subgroup-based multicast strongly depends on the virtual group structure used for local error recovery and acknowledgment handling. The paper does not define an exact algorithm for subgrouping multicast receivers, but presents some guidelines to do so:

- *Small subgroups should be avoided in flat hierarchies*: In small subgroups, the probability of being able to recover from errors by local retransmissions is quite low. The more members a subgroup includes, the higher the probability that one of them correctly receives a certain data packet. Instead of introducing overhead for management of small subgroups, it is preferable to combine them into a larger group managed by just a single controller. Nevertheless, the size of the subgroups must not cross a certain threshold due to controller implosion.
- *Multiple subgroups should be arranged in a multi-level hierarchy*: The arrangement of multiple subgroups into a multi-level hierarchy relieves the

sender from acknowledgment processing and reduces the number of retransmissions performed by the sender. While the establishment of multi-level hierarchies comes at the expense of additional complexity, the benefits in multicast performance will justify it.

- *Bad receivers should be fairly distributed among all the subgroups*: A single subgroup that permanently requests missing data packets from the sender degrades overall multicast performance. Therefore, multicast receivers should be arranged in a way that each subgroup includes at least one receiver with low packet loss rate.

The results exhibit the importance of packet loss rate for the establishment of virtual group structures. Rather than solely using a hop count metric, the packet loss rate should be taken into account for the arrangement of receivers into subgroups.

Currently, further experiments are performed in the MBone to find an optimal setting for the various parameters of DCP. An implementation of DCP as well as of LGMP is available for Digital Unix, SunOS and Linux. More information on the project, related research activities and future work could be found at http://www.telematik.informatik.uni-karlsruhe.de/~hofmann/LocalGroups.html.

References

1. S. Deering: *Host Extensions for IP Multicasting*. Internet Request for Comments RFC 1112, August 1989.
2. C. Diot, W. Dabbous, J. Crowcroft: *Multipoint Communication: A Survey of Protocols, Functions, and Mechanisms*. IEEE Journal on Selected Areas in Communications, Vol. 15, No. 3, Pages 277-290, April 1997.
3. S. Floyd, V. Jacobson, S. McCanne, C.-G. Liu, L. Zhang: *A Reliable Multicast Framework for Light-weight Sessions and Application Level Framing*. Computer Communication Review, Vol. 25, No. 4, Proc. of ACM SIGCOMM'95, August 1995.
4. M. Hofmann: *A Generic Concept for Large-Scale Multicast*. International Zurich Seminar on Digital Communication, February 21-23, 1996, Zurich, Switzerland, Ed.: B. Plattner, Lecture Notes in Computer Science, No. 1044, Pages 95-106, Springer Verlag, 1996.
5. M. Hofmann: *Enabling Group Communication in Global Networks*. Proceedings of Global Networking'97, Volume II, Pages 321-330, Calgary, Alberta, Canada, June 1997.
6. V. Kumar: *MBone - Interactive Multimedia on the Internet*. New Riders Publishing, Indianapolis, Indiana, USA, 1996.
7. S. Paul, K.K. Sabnani, J.C.-H. Lin, S. Bhattacharyya: *Reliable Multicast Transport Protocol (RMTP)*. IEEE Journal on Selected Areas in Communications, Vol. 15, No. 3, Pages 407-421, April 1997.
8. M. Yajnik, J. Kurose, D. Towsley: *Packet Loss Correlation in the MBone Multicast Network*. IEEE Global Internet '96, London, England, November 20-21, 1996.
9. R. Yavatkar, J. Griffioen, M. Sudan: *A Reliable Protocol for Interactive Collaboration Applications*. Proceedings of ACM Multimedia'95, 1995.

Springer
and the
environment

At Springer we firmly believe that an
international science publisher has a
special obligation to the environment,
and our corporate policies consistently
reflect this conviction.
We also expect our business partners –
paper mills, printers, packaging
manufacturers, etc. – to commit
themselves to using materials and
production processes that do not harm
the environment. The paper in this
book is made from low- or no-chlorine
pulp and is acid free, in conformance
with international standards for paper
permanency.

Lecture Notes in Computer Science

For information about Vols. 1–1278

please contact your bookseller or Springer-Verlag